Modern Toys From Japan
1940s – 1980s

William C. Gallagher

4880 Lower Valley Road, Atglen, PA 19310 USA

Many of the toys or the prototypes from which they were derived may be covered by various copyrights, trademarks, and logotypes. Their use herein is for identification purposes only. All rights are reserved by their respective owners.

The text and items pictured in this book are from the collections of the Masudaya Corporation, the author, or other various collectors and collections. This book is derived from the author's research along with material provided by the Masudaya Corporation and discussions with Masudaya employees.

Copyright © 2005 by William C. Gallagher
Library of Congress Control Number: 2005928956

All rights reserved. No part of this work may be reproduced or used in any form or by any means—graphic, electronic, or mechanical, including photocopying or information storage and retrieval systems—without written permission from the publisher.

The scanning, uploading and distribution of this book or any part thereof via the Internet or via any other means without the permission of the publisher is illegal and punishable by law. Please purchase only authorized editions and do not participate in or encourage the electronic piracy of copyrighted materials.

"Schiffer," "Schiffer Publishing Ltd. & Design," and the "Design of pen and ink well" are registered trademarks of Schiffer Publishing Ltd.

Designed by Mark David Bowyer
Type set in Geometric Hv BT/Humanist 521 BT

ISBN: 0-7643-2313-X
Printed in China
1 2 3 4

Published by Schiffer Publishing Ltd.
4880 Lower Valley Road
Atglen, PA 19310
Phone: (610) 593-1777; Fax: (610) 593-2002
E-mail: Info@schifferbooks.com

For the largest selection of fine reference books on this and related subjects, please visit our web site at
www.schifferbooks.com
We are always looking for people to write books on new and related subjects. If you have an idea for a book please contact us at the above address.

This book may be purchased from the publisher.
Include $3.95 for shipping.
Please try your bookstore first.
You may write for a free catalog.

In Europe, Schiffer books are distributed by
Bushwood Books
6 Marksbury Ave.
Kew Gardens
Surrey TW9 4JF England
Phone: 44 (0) 20 8392-8585; Fax: 44 (0) 20 8392-9876
E-mail: info@bushwoodbooks.co.uk
Free postage in the U.K., Europe; air mail at cost.

Contents

Acknowledgments 4
Foreword by Hank Saito 5
Introduction 6
Historical Overview 8
    The Masutoku Toy Factory 9
    The Modern Toy Laboratory 9
    Toy Materials 9
    Key Developments 10
    The Role of Importers 13
    Masudaya Marks 15
    Important Milestones 16
What's in This Book? 18
Scarcity, Condition and Values 19
Aircraft 21
    Propeller Airplanes 21
    Jet Airplanes 24
    Military Airplanes 28
    Aero-Mini/Mini Air Series 30
    Helicopters 33
Amusement Park &
    Kiddy Vehicle Toys 39
    Amusement Park Rides 39
    Kiddy Vehicles 42
Animals 50
    Bears 50
    Birds 55
    Bugs & Butterflies 58
    Cats 59
    Dogs 62
    Donkeys 66
    Elephants 66
    Monkeys 67
    Rabbits 69
    Turtles 70
    Other and Grouped Animals 70
Boats 73
    Motorboats 73
    Cabin Cruisers 75

Ships 76
Schooners 78
Paddlewheel 78
Tugboats 79
Submarines 80
Buses 81
Cars 85
    Old Timers 85
    Passenger Cars 86
    Assembly Kits 96
    Trailers 97
    Emergency Cars 97
    Racing Cars 105
    Platform Base Toys 109
Character Toys 110
    Disney Toys 110
    Japanese Characters 122
    Tom & Jerry 136
    Other Character Toys 140
Circus & Clowns 142
    Circus Animals 142
    Clowns 143
Farm Tractors 146
Games 147
House Play Toys 150
    Appliances and Cookware 150
    Baby Toys 155
    Dolls 156
    Miniatures 158
    Phones 158
Military 160
    Armored Cars 160
    Cannons & Guns 160
    Jeeps 163
    Tanks 164
    Trucks 170
    Other 170
Motorcycles 171

People 175
Space 181
    Ray Guns 181
    Robots 183
    Space Exploration and Travel 187
    Plasman 207
Trains 209
    Animal Trains 209
    Diesel Trains 211
    Electric Trains 213
    Platform Base Trains 214
    Steam Trains 215
    Trolleys 226
    1776 USA Bicentennial Trains 227
    Train Sets 228
Trucks & Construction 232
    Trucks 232
    Fire Trucks 234
    Bulldozers 237
    Cranes 237
    Forklifts 238
    Road Rollers 239
Other Collectible Toys 240
    Miscellaneous 240
    Comic Toys 241
    Counter Toys 242
    Gobblers 244
    Pullback Toys 246
    Walkers 246
    Water Toys 250
Ads & Catalogs 251
Bibliography & Resources 256

Acknowledgments

As with my previous books, I am continually grateful to the many people who willingly give of their time to help with my research. This includes everything from gathering historical data; to sharing toys, knowledge, and photographs; to help with the Japanese language. Without these contributions, a reference book of this nature would not be possible.

First, the experience, knowledge, and documentation shared by **Hank Saito**, Director, and **Harumasa Saito**, President and Owner of the Masudaya Corporation became the foundation for this effort. From this base of information, many others provided support in the many ways mentioned above. My sincere thanks to all!

Rodney Abensur, Tom Allen, Rex & Kathy Barrett, Bertoia Auctions, Don Bryant, Donald Conner, Rich Finney, Les Fish, Brian Flynn, Lionel & Annie Fournier, Sonny Glasbrenner, Carl Gruber, Hakone Toy Museum, Dan Henry, Ayako Honda, Don Hultzman, Masamichi Inoue, Japan Toys Museum, Steven Jaspen, Robert Johnson, Eiji Kaminaga, Dean Klein, Stan Luksenburg, Mikio Machii, Joe Morabito, Barbara Moran, Morphy Auctions, Shigeru Mozuka, Mark Nagata, Ike Ogawa, Justin Pinchot, Ray Rohr, Harumasa Saito, Barry Skelly, Jack Smith, Ron Smith, Smith House Toys, Bruce Sterling, Super 7 Magazine, Sumihiko Takasawa, Toyoji Takayama, Yoshio Udagawa, Kikuo & Kazuyo Uno, Kozo Wakabayashi, Tosh Wakabayashi, Larry Waldman, Barry Weand, Bill & Stevie Weart.

Of course, I also want to thank my publisher and specifically my editor, **Donna Baker** who has now helped me through four books with her skill and guidance.

Foreword

The Saito Family and the Origin of Masudaya Corporation

by Hank Saito, Director

The founders of the Masudaya Toy Company trace their ancestry back to Fuku Saito, who was Kasuga-no-Tsubone in the Edo castle during the Tokugawa period of Japanese history. Fuku Saito was the famous governess, tutor, and private wet nurse of the Third Shogun, Iemitsu.

Rinbei Saito (the fourth generation of Fuku Saito's third son) decided to give up the Samurai Caste Class and become a merchant, as he loved the free and easy life. Rinbei Saito opened a small toy store in 1724 in Okuyama, Asakusa, a large shopping district in downtown Edo that is now Tokyo today. It was eighty years after the death of Fuku Saito. The graves of the Saito family, owners of Masudaya, are in the Rinsho-in Temple yard of the Bunkyo Ward in Tokyo.

In 1724, Rinbei Saito began selling hand crafted Japanese dolls and folk craft toys using natural materials of wood, paper, cloth, and clay. Two hundred and eighty years later, Harumasa Saito, an eighth generation descendent of Rinbei Saito, is now the current president and owner of Masudaya Corporation (Masudaya Saito Bokei), Japan's oldest operating toy company.

Though the Tokugawa reign lasted almost three hundred years, in 1868 rule was surrendered to Emperor Meiji. He was regarded as a great emperor, overseeing the country during a period of great reform, modernization, and the restoration of imperial rule. This period of 1868 to 1912, referred to as the Meiji Restoration period, saw revolutionary changes in both Japanese culture and the structure of society.

Japan became a modern, industrialized country very quickly and the toy industry was no exception. Manufacturers made more modern toys, utilizing various materials and covering a variety of subjects. Japanese toys were becoming accepted worldwide and by the late 1920s and early 1930s, Japan had become one of the world's significant toy producing countries.

However, World War II destroyed the toy industry, requiring it to start over from the beginning. Some companies never recovered. Masudaya did start over and dealt with the material shortages. The lack of tinplate was a big problem and many factories resorted to using old tin cans, such as those from beer or soft drinks. These cans were flattened with hammers until presses were once again available. Occupied Japan toys were labeled accordingly through 1952 and are still sought by enthusiastic collectors.

During the 1950s through the early 1970s, toys again became one of Japan's biggest exports, as this period saw Japanese toys gaining in popularity and being exported around the world. As Japan prospered and labor costs increased, other developing countries with lower costs became the primary source for toys by the late 1970s. Production continued to shift to these developing countries and today almost ninety percent of toys are manufactured in China.

Harumasa Saito, President, Masudaya Corporation.

Introduction

Masudaya Corporation – The Modern Toy Company

Hank Saito, Director and 50+ year employee of Masudaya, has been the primary contributor to the documentation and understanding of the Masudaya Corporation and the workings of the Japanese toy industry during their post-war "Golden Years." (While bearing the same family name as the founding family of Masudaya, he is not related.) Having joined the company in 1953 as a young graduate with English language skills, Mr. Saito witnessed firsthand the phenomenal growth of the Japanese toy industry. With his language skills, he became Masudaya's focal point for dealing with English speaking buyers. Toy collectors and historians are both indebted to Mr. Saito for his willingness to share his experiences and knowledge.

Speaking of his early years, Mr. Saito notes: "I was born in Yokohama. My father was working for a German company in Yokohama that was at one time the largest customer for Masudaya. My home burned down during the war and I had many rare toys that the owner of Masudaya had handmade for me, but they are all gone. At that time, German toys were the originals that everybody copied, then changed the style and color. That was the beginning of the growth in the Japanese toy industry."

Due to many complaints about quality, in 1953, Masudaya's president Haruhiro Saito (grandfather of the current Masudaya president) went to the United States with Hank Saito as his secretary and interpreter. He was concerned about the reputation of Japanese toys. The government was also concerned about export quality. The president of Masudaya, along with Mr. Isshi of the Alps Toy Company, Mr. Nomura of Nomura Toy, and Mr. Yonezawa of Yonezawa Toys—representing the four biggest Japanese toy companies—planned to set up a private inspection organization to promote and control quality. They established the Japan Toy Manufacturing Wholesalers Association. Haruhiro Saito served as the Association's first Chairman and was a great contributor to the quality improvement and growth of Japan's toy industry. He recognized the importance of quality and invited government managers including MITI, Japan's Ministry of International Trade and Industry, to join the association, which continued to exist for nearly fifty years. Shipments could not be made without the inspection certificate issued by the testing laboratory of the Association. In recent years it became somewhat meaningless, since exports were low. The Inspection Association eventually became the Japan Toy Association.

The first Japan Toy Fair was held in 1962. At that time, the toy fair was not export oriented. It focused mainly on domestic requirements. Foreign buyers would occasion-

Hank Saito, Director, Masudaya Corporation.

1962 Masudaya Toy Fair ad.

ally come, but they did not really care about the toy fair since they did not want their competitors or the public to see their prospective merchandise. Also, most already knew how to contact and work with the Japanese toy manufacturers.

I have seen many references in the US indicating that Masudaya was started in 1924, but these are in error. As you read in the Foreword, 1724 is the correct date. Still operating in 2005 earns Masudaya the distinction of being Japan's oldest operating toy company. In today's era of consolidation and global competition, it is hard to imagine a toy company that has experienced its 281st birthday.

Masudaya is a full line toy company, having sold everything from baby toys and general play toys to model kits. In the US, we know them most for their play toys imported into the US from the late 1940s to the 1980s. These peak years of export are often referred to as the "Golden Years." It is this period that the book focuses on primarily, with a few other toys included for their general interest. You will find a large number of popular toys of the era referenced in this book.

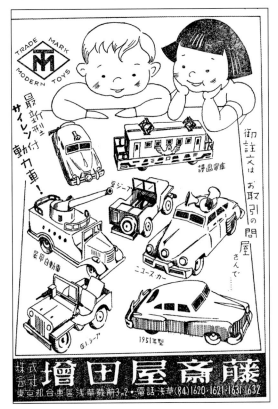

Trade journal ad, October 1950.

Trade journal ad, November 1952.

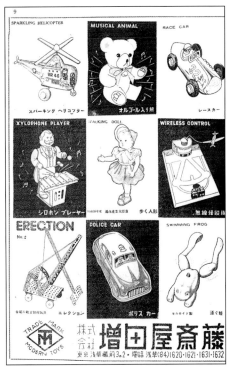

Trade journal ad, April 1952.

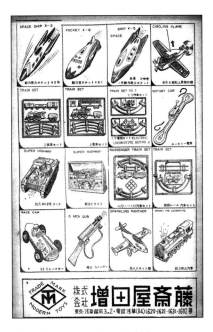

Trade journal ad, August 1952.

Historical Overview

Since most of this book focuses on toys produced by Masudaya during the period when Japan was the largest toy producer, it is important to understand the structure of the Japanese toy industry. The Masudaya Toy Company and its subcontract partners are typical of this structure.

Where are the big toy factories?

Early importer buyers came to Japan looking for the big toy factories only to find out they did not exist. The manufacturing system in Japan was much different than in other countries. You often did not see the real manufacturer because of Japan's unique approach to production. For example, Nikon, the famous camera company, at one time made only fifty percent of their camera components. They concentrated on the most important parts, such as the lens or shutter, and subcontractors made the remainder of the parts. Likewise, the toy makers also depended heavily on the use of subcontractors.

Many of the toy companies relied exclusively on the use of subcontractors, so talking about toy makers can be very complicated. Here is a surprising fact: The three largest toy companies in the 1950s were Masudaya, Nomura, and Alps and yet none of these companies ever had their own factory! In fact, most of the Japanese toy company names we are familiar with did not have their own factories.

For most of their toys, Masudaya used three or four major subcontractors, who in turn used other subcontractors. When Masudaya visited their subcontract stamping factory, they would also see toy stampings for other toy companies. Initially, no single subcontract toy factory owned a 100-ton press, which was used for larger pieces. So the toy factories would go to non-toy industry stamping press companies for these large pieces.

With tin toys, it was very difficult to make figures; Masudaya would make the mock-up and the factory would modify the design. Sharp angles could be pressed by hand, but the factories would have to modify the tooling for a moderate slopes or curves. The factories therefore had their own technology and capabilities. Surprisingly, the technology for pressing tin toys was very advanced. Hank Saito still remembers one of the very old manufacturers that made the bumpers. Distributing even power across the piece during stamping is very important and for shapes like bumpers, it was very difficult. Masudaya's subcontractor made it with one piece in such a way that it could be cut later after pressing. Interestingly, Japan's real auto manufacturer, Prince Skyline (now Nissan), was surprised to learn about this approach and studied how it might be adapted for their applications. So the stamping presswork was very dedicated and a high quality of work was required. Foot presses were used for small pieces, but up to 100-ton high power, motorized presses were usually required for the large pieces used in large toys and trucks.

Work done at people's homes was very important to keeping the cost down. A coordinator would distribute parts, including tin sheets and even small foot presses, to individual homes, and the following day assembled parts or stamped parts would be picked up. As Korea, Taiwan, and Hong Kong developed, they often followed a similar practice. Even though work was done at home, it was uncommon to see child labor in Japan.

Typically, the subcontract stamping or pressing factories would also do the final assembly but would go to other companies for parts. Even the stamping dies or molds came from specialists. Most of the industry would go to the same companies. For example, rubber tires came from one or two manufacturers making only tires, from small sizes to large sizes. Two or three manufacturers made the clockwork spring motors. Mabuchi, the electric motor maker established in 1954, was used by virtually everyone making battery-operated toys. Other toy motor makers included Mizuno, who is credited with producing the first electric motorized toy in Japan around 1952 or 1953, and Kato Sairen (Nihon Boeiki). Gears and cams also came from two or three companies, but the assembling factories usually produced the actual gearboxes, due to their many variations.

Subcontractors' marks were typically not seen on Masudaya toys. Masudaya would normally design their own toys, apply for patents, pay for the molds and materials, and not allow the subcontractor to use its own name or mark on the toy. It is unusual to find a subcontract trademark on a Masudaya toy. If a subcontract mark is seen on a toy, it is likely that the subcontract factory owned the patent or design and tooling instead of Masudaya. Note: This should not be confused with the mark of an importer or trading company, which will be discussed later.

The Masutoku Toy Factory

Many references to the K.K. Masutoku Toy Factory are found when researching Masudaya. Some writers have even suggested that Masutoku was the Japanese company name behind Modern Toys. If Masudaya did not own factories, what is the connection to the Masutoku factory that is mentioned in some Masudaya historical catalogs?

There was a time during which the US imposed an import duty on selling prices that were ex-factory, meaning not from a factory. As a result, Masudaya customers and importers wanted to buy direct from the factory. Since there was no factory, the customers asked that Masudaya establish a factory to avoid this duty.

Masudaya decided to use the name Masutoku for a non-existent manufacturing factory. Masutoku was the cable address for Masudaya back in the 1920s. At that time, the given name of the Masudaya family head was Tokutaro, so the cable name chosen was Masutoku—a combination of both names. The intent was to provide the company's customers with an exemption from an import duty of about 10%. This was done around 1949. Driven by US customer demand, this became a common practice for Japanese toy companies, resulting in most companies using two names. While the levy of this type of import duty showed a lack of understanding by the US regarding manufacturing in Japan, the practice was ultimately discovered and the industry quit using fictitious factory names, as the US government would not allow it.

The Modern Toy Laboratory

The initials MT or TM seen in Masudaya's logo and trademark stand for Masudaya Toys, even though the inclusion of the words "Modern Toy" in their trademark leads people to believe that MT stands for Modern Toy.

In the 1930s, Masudaya established a "Modern Toy Laboratory" within their facility. The purpose of this laboratory was to plan, design, and hand build sample "modern" toys for potential production. The Modern Toy Laboratory concept was promoted heavily by emphasizing Masudaya's capability in modern toy design. The popularity eventually led to the inclusion of the words "Modern Toy" in their trademark.

To develop or produce a toy, a Masudaya designer would make a sketch, (not a blueprint, just a sketch) and based on that sketch a handmade sample, called a mock-up, would be made. The engineers usually made working samples. About ten people working in the Modern Toy Laboratory were involved in making these handmade samples. The mock-ups were very expensive, sometimes costing as much as $5000 to build, due to the amount of work involved. From these samples, cost estimates were made that would be presented to customers in an effort to obtain orders.

The Modern Toy Laboratory was responsible for many technical developments and adaptations. The Radicon line, the Sonicon line, and the whistling locomotives were all products of the laboratory. Adapting other toy technology such as electric motors, mystery action, and non-fall action were all part of the group's responsibility.

Toy Materials

Masudaya toys made before World War II can be found in celluloid and/or tin. After the war, both materials became problems. Celluloid tended to crack over time and burned easily. The flammability of celluloid (cellulose nitrate) made it dangerous to manufacture, so after the war, non-flammable celluloid type toys (cellulose acetate) were produced. However, they were not as popular as tin toys. Original celluloid toys were very easy to color, but cellulose acetate was more difficult to color. Most post-war toys identified as celluloid are cellulose acetate.

Tin was not readily available immediately after the war, resulting in the re-use of old tin cans. It is not uncommon to find toys that still show evidence of having had a previous life as a tin can. This can be observed by looking at the interior of the tin where original can printing can be found. The cans were flattened, then either painted or printed (lithographed) before being pressed and stamped into a different shape or size.

As tin availability increased, specialized stamping factories emerged to serve the toy industry. Lithography processes also improved and bright colorful toys from Japan appeared in stores around the world. The color and designs led to increased popularity and eventually to becoming desired collectibles. One of the benefits of tin was the ability to use the dies over and over again. By just changing the lithography, the same dies could now produce a multitude of toy variations. Passenger cars could become emergency vehicles. Military jeeps could become circus jeeps. Military tanks became space tanks, Western locomotives became old-fashioned locomotives and a Santa Fe train

10   Historical Overview

became a Union Pacific train, all from the same shaped parts. A passenger bus could be turned into a character bus just by changing the lithography, without changing the shape or design. As you look through this book, you will see many examples of different toys produced from the same tooling or molds.

Figures were more difficult to make so the availability of vinyl allowed tin heads to be replaced with vinyl heads. A change in a vinyl head could turn a normal toy into a Disney toy, a Tom & Jerry toy or an Ultraman toy.

Plastic was a less expensive material and the high technical precision required for tin stamping was not easily obtained. While plastic did not lend itself to all of the things mentioned above, the overall costs were still less and a gradual shift from tin toys to plastic toys was observed during the 1960s and 1970s.

Today, it is difficult to make tin toys in Japan. The sheet lithograph process is essentially obsolete. Now most printing (and stamping) is accomplished utilizing higher volume roll fed presses, the same as can makers use, instead of the sheet fed printing that was originally employed in the toy industry.

Key Developments

Mystery Action Toys

The mystery action concept originated from a product, patented in the United States, called *Marvelous Mike*. This robot driven bulldozer incorporated a set of drive wheels that were free to spin on a center axis. If the toy hit an obstacle or a wall, this set of drive wheels would turn and the toy would move forward in a different direction. This action could be repeated indefinitely, making the toy seem very mysterious in its actions. Masudaya incorporated a version of this in many, many toys because of its great popularity. It was sometimes called the non-stop action, but Masudaya primarily used the term "Mystery Action." Many toy collectors also refer to it as bump and go action.

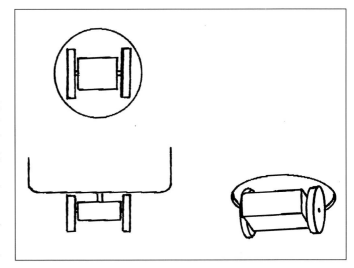

Mystery action or non-stop action illustration.

Non-Fall Action Toys

The non-fall mechanism utilized in Masudaya toys was the creation of Mr. Wada of the Towa Company. Originally, Schuco of Germany made non-fall toys starting back in the 1920s with a wheel that would stop the mechanism if the toy went over an edge. Mr. Wada improved that mechanism with a special bar that was more effective, but the non-fall mechanism was only suitable for longer toys due to the fact that the center of gravity must always be in the rear part of the toy. Non-fall toys were some of the best selling items produced by Masudaya.

When such a toy begins to go over the edge of a table (from any angle while moving forward), the non-fall bar drops, engaging a drive wheel that is perpendicular to the forward direction main drive wheels. When this perpendicular drive wheel engages the table surface it causes the toy to change direction away from the edge. Accordingly, the toy does not fall from the table and the name "non-fall" is applicable.

Trade journal ad, October 1956.

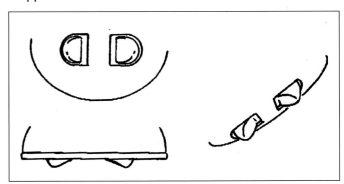

Non-fall action illustration.

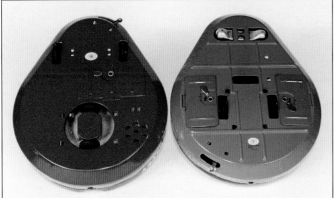

Photographs of similar toys produced in both mystery action and non-fall action versions.

Radicon©, The Radio Control Technology

As early as 1953, Masudaya began discussing the concept of the world's first radio control toy and began their research and development. After spending almost two years in development, the world's first radio remote controlled toy was completed in 1955. Masudaya introduced a toy containing a system that was very unique but obsolete by today's standards. Based on a system developed by Marconi, the famous Italian physicist and Nobel Prize laureate, the radio control system uses a Coherer tube as a wave detector, which is one of the oldest forms of radio transmission. The transmitter had no particular frequency, essentially emitting various and miscellaneous frequencies. Control was in sequential steps—the first push of the button caused the toy to go straight, pushing again caused the toy to turn right, the next push made it go straight again, then left, then straight again and then stop with each subsequent push of the button. The system was first put into a bus body and named *Radicon Bus*. Radicon was an abbreviation of "radio control" and was registered as a copyright worldwide.

Press Action Toys / Lever Action Toys

Press Action toys were popular in the early 1950s. The concept was to push down at a certain point on the toy and release to watch the toy move forward. This was accomplished by the use of a spring and a conventional inertia friction motor. So instead of pushing the toy to activate the friction motor, you pressed down on the toy. The idea was expanded to utilize other ways of activating the mechanism, including the trolley poles on a trolley or an antenna lever on a car. The concept later became the basis of the lever action toys.

Press action toys shown in trade journal ad, June 1953.

Trade journal ad, December 1955.

12 Historical Overview

Following the bus, Masudaya wanted to install this system in a robot. The *Radicon Robot*, introduced in 1957, was made of relatively thick tinplate that required a heavy duty stamping machine and specialized tooling. Masudaya succeeded and the first robot in the "Gang of Five" series was born.

The *Radicon Boat* was also introduced in 1957. Over the next ten to fifteen years, a variety of Radicon toys were introduced, including a Mercedes Benz, Missile Tank, Bullet Train, Space Path Finder, Police Patrol Car, Western Type Locomotive, and Airplane. In the late 1970s the Radicon name was adopted for a series of multi-channel radio control toys.

After extensive study and development, Masudaya successfully minimized the size of this device and incorporated it into the locomotive itself. This was a great achievement for Masudaya. The device consisted of a motorized hooting drum combined with a moving valve to create the echo sound as if the locomotive were actually running with a very realistic whistling sound. The toys that resulted from this development were among Masudaya's greatest successes, resulting in many different toys for over twenty years.

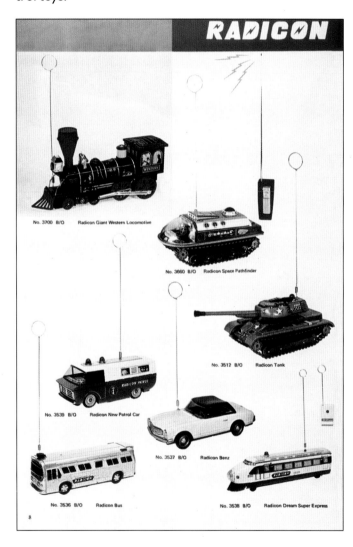

Sound Control Toys

In 1957, Masudaya introduced their first sound control toy, the *Sonicon Rocket*. A whistle was included with the toy. Each toot on the whistle would cause the battery-operated toy to change directions. Sonicon technology was also used in buses and cars from its introduction through the early 1970s.

Floating Toys

The use of a motor driven blower to suspend a ball in the air over a toy began in the late 1950s. A ball blowing train, car, fire engine, and space tank were introduced in 1959. The ball was made of Styrofoam. For space toys, the ball made a great satellite and many space vehicles were soon introduced with floating "satellites." Masudaya continued to refine this concept and soon introduced paper astronaut figures mounted in a Styrofoam half shell designed to let the astronauts float over the space capsule as they "walked in space."

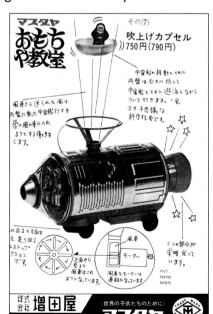

Trade journal ad from December 1970 illustrating *Space Capsule with Floating Astronaut* and the internal blower mechanism.

Hooting Whistle

In late the 1950s and early 1960s, the device that simulated the whistling and hooting of a locomotive was sold separately because of its large size. Toy train makers often put the device into a building structure such as a station or platform and marketed it as one of the accessories of a train set.

The Role of Importers

Initially, when the toy business was small, Masudaya's major customers were importers. One of the largest was New York Merchandise (NYMCO). Others included F.J. Strauss (STRACO), Illfelder (later ILLCO), IB Cohen & Sons (SONSCO), George Wagner (later WACO), George Borgfeldt Corporation, and M. Pressner. Companies such as Cragstan, Linemar, Rosko, AHI, Frankonia, and Shackman followed them later in the mid 1950s.

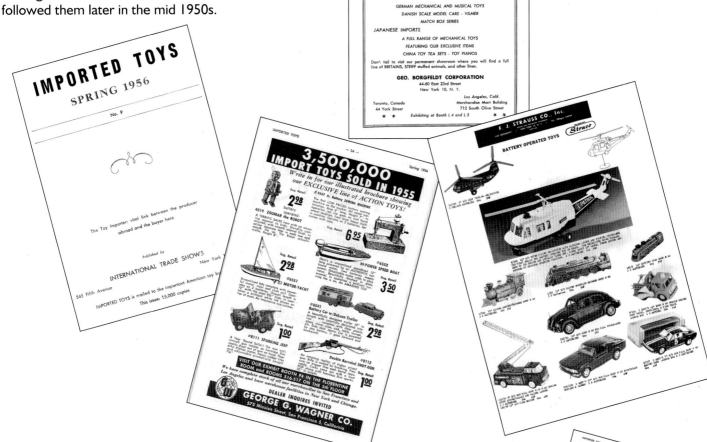

A typical order from an importer would be around 120 dozen or 1440 pieces. Orders of 360 dozen (4320 pieces) were considered big orders. Consider the scarcity today if only 1440 pieces were ever made! Sometimes the importer would put their own trademark or name on the box, but not on the toy.

When the US importer Cragstan came to Japan in the mid-1950s, they had been importing automobiles and did not have a toy background. They were very aggressive and started asking for pricing for 1000 dozen or sometimes even 2000 or 3000 dozen. They asked for exclusive distribution and for toys bearing their own name. Previous importers had never asked for their own name. This represented a significant change in the industry, both for pricing and for manufacturing. Masudaya produced many toys for Cragstan over time.

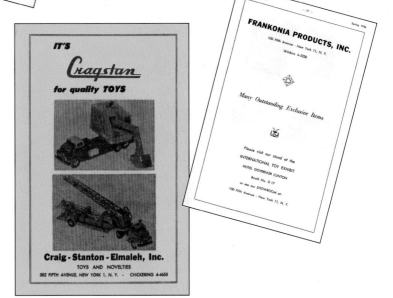

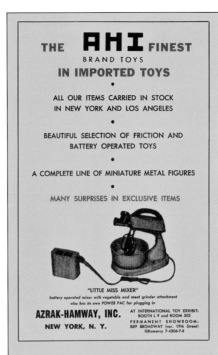

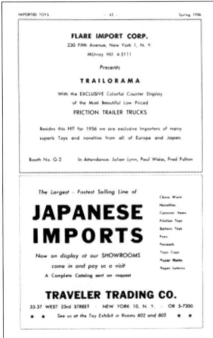

Here is an example of how private label toys evolved: Cragstan would come to Masudaya in Japan and order 3000 dozen of a certain style or color, then AHI (US importer Azrak-Hamway Inc.) would also want a certain color but it might have already been sold to Cragstan, forcing AHI to choose a different color. These toys would be shipped from Japan in private label boxes but labeling laws still required them to be marked "Made in Japan." This private label practice resulted in many variations of the same basic toy. Sometimes color was the variation, as just mentioned; sometimes different features, such as bells and whistles on trains, were added. Often, however, the variation was just a different toy name in a different box. This practice can cause confusion amongst collectors if not understood.

Marx established their import office in Japan as Linemar (Marx Line). They were already in the toy business and also asked for their own name or private label. Marx often had the licensing rights for many character toys. Even long established Illfelder asked for their own name.

In effect, these companies would give the impression that they were the actual manufacturers. Some of the independent importers were backed financially by the big retailers, not in ownership, but in guarantee of payment. Toys made with another company's private label name or by special order for that company, were often referred to as OEM (Original Equipment Maker or Manufacturer) variations.

During the 1950s and 1960s, the US was the largest destination for Masudaya toys, representing perhaps 70% of their production. Another 20% went to Europe and the balance to the rest of the world.

It should be noted that Masudaya also acted as a trading company for other toy companies in Japan. This allowed consolidated shipments to be made to the importers.

From Idea to Customer

With so many toys having been produced in Japan, we often wonder about the toy production cycle—from concept to the finished product. After WWII, as the Japanese toy industry began to flourish again, the toy manufacturers first produced typical toys that had been popular in the past. But toy importers began to request specific toys and often provided a German or American toy for the toy makers to copy. Even though the toy subsequently produced was a copy of or similar to an existing toy, the market was driven by what the customers wanted. And the customers were mainly US importers.

Basically, there was a kind of rotation every year. The buyers would come in March or April with ideas. Masudaya would then design the toy and make a model. The buyers returned in the month of September and were presented with the mock-up samples and pricing. Trial orders were then placed on approved items. It is interesting to note that the US importers would also shop the toys around, looking for other toy makers to produce items at a lower price.

When auto manufacturers announced a new car model, maybe six or seven toy manufacturers would make a toy based on that new model. Often, however, the toy production cycle did not match the real auto cycle of introducing new models. This would result in toy makers producing "hybrid" models carrying characteristics of two (or more) different model years for the real autos.

Once the decision was made to make a product, Masudaya would start producing the molds or dies through their subcontractors. At the same time, they often duplicated the mock-up samples to be presented by the importers at the New York Toy Fair in February.

In October or November of each year, the major retailers like Woolworth, S.S. Kresge/K-Mart, Sears, J.C. Penney, Montgomery Ward, and others would come to Japan and place final orders. Some retailers dealt only through a single importer, which required Masudaya to sell to that importing company, but some chain stores came directly to Masudaya requesting that Masudaya export directly to that retail store chain. (Again, Masudaya accepted orders given to Masutoku as the factory.) Eventually, the large retailers' names began to show up on the toy boxes.

Still other wholesalers, like Pensick & Gordon, Lachman Rose, Greeman Brothers, etc., would come in December or January to place orders for already designed toys. From after the Toy Fair until August, Masudaya would concentrate on manufacturing.

The importers would place repeat orders in the April to June time frame, for shipments starting in April/May and reaching a peak in July or August. For the east coast, the end of August was the shipping deadline for making the Christmas season. For the west coast, the shipping deadline was September 15th.

Masudaya would also introduce toys based on previously manufactured toys. By adding new functions to an existing model, such as a light or bell, a new toy could be introduced. This was an advantage of tin that was not possible with plastic toys. For example, one of Masudaya's largest selling toys was the tin locomotive. Over two hundred different locomotives were derived from the original style.

Masudaya Marks

The trademarks used over the years by Masudaya were pretty consistent or similar since their introduction. Starting with the fundamental mark of MT for Masudaya Toys to the addition of the words Modern Toys (from their Modern Toy laboratory), the MT for Masudaya Toys has remained the same, making identification relatively easy. There were toys produced with no marks and occasionally with dual marks, but those are the exceptions.

Early mark (used on toys through the 1930s and boxes pre 1920s).

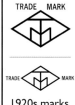

1920s marks as seen on boxes.

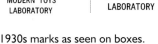

1930s marks as seen on boxes.

1940s marks as seen on boxes (Made in Occupied Japan).

1950 – 1992 marks.

Current marks.

Important Milestones

1724 Origin of Masudaya.

1904 Started handling toys in addition to dolls and folk craft.

1925 Changed to a company organization from private toy store.

1926 Company moved to current location and began export business.

1940 Toy exports to America and Europe are stopped.

1941 Operation stopped due to the war.

1946 Business operation resumed.

1948 *Dancing Couple* goes on sale and is a hit in the export market.

1948 Friction motors are applied to automobile toys.

1949 *Walking Bear* is introduced and becomes an epoch hit in the export market.

1951 Walking *Jolly Penguin* introduced and receives award from Ministry of International Trade and Industry (MITI).

1952 *Walking Doll* goes on sale and receives MITI award.

1952 The first battery-operated electric motors are applied to toys.

1955 The first wireless control Radicon toy *Radicon Bus* goes on sale.

1957 The *Radicon Robot* is introduced becoming the first of the "Gang of Five" robots.

1957 Sound control toy is developed and the *Sonicon Rocket* is introduced as the first toy with the Sonicon technology.

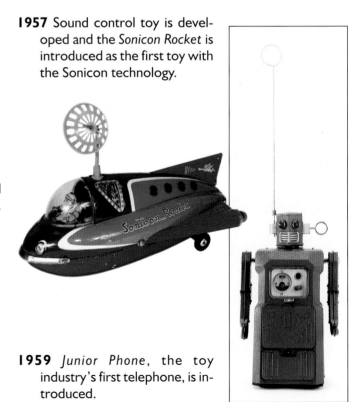

1959 *Junior Phone*, the toy industry's first telephone, is introduced.

1960 *Flying Saucer X-7* goes on sale. Space flying saucers based on this design were produced for over twenty years.

1962 First Japanese International Toy Fair held in Tokyo.

1962 *Western Special Locomotive* is introduced. This toy was awarded the top Gold Prize for export success.

1963 *Astroboy (Tetuwan Atom)* is introduced on TV and sets the stage for Japanese character based toys.

1964 Mr. Shigeru Saito becomes the 7th president of Masudaya (father of current president).

1964 *Old Fashioned Fire Engine* is introduced and receives Silver Prize.

1965 *MS-58 Missile Tank* is introduced.

1966 Japan is largest toy exporter in the world.

1967 *Silver Mountain Locomotive* is released and is awarded Gold Prize.

1968 *Overland Express* goes on sale and is awarded Diamond Prize at toy fair.

1972 *Missile Robot MR-45* is introduced and later receives Diamond Prize at Toy Fair.

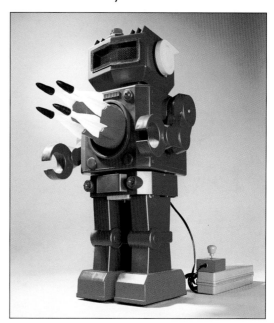

1972 Japanese toy industry adapts "Safe Toy" standards and marking system. The corresponding **ST** mark is applied to toy packaging from this time forward.

1972 Hong Kong becomes the world's leading toy exporter.

What's in This Book?

With Masudaya having produced general line toys and children's playthings for over 280 years, one could fill volumes with pictures and illustrations of the company's toys—if that information were available. Unfortunately, however, most of the prewar history was lost during World War II. As a result, this book primarily covers the "Golden Years" of Masudaya, when toy production was at its highest. These "Golden Years" are represented by toys from the 1940s through the 1970s. Some examples of toys produced both before and after that period are also included, to offer a broader view of Masudaya's extensive line of toys. Currently, toys made between 1946 and 1980 can still be found on the antique toy and collectible market and are the ones sought by most collectors.

In an effort to make it easier to find a specific toy in this book even if the toy name is not known, toys have been arranged by collectible subject category. These categories and the subcategories within them are listed in the Contents. Please keep in mind that there were often variations in the name of a specific toy and even the catalog name might be different than the final name on the box. This was usually the result of the customer requesting a specific name, private labeling, or a name they felt was more suitable for their markets. In addition, many toys obviously cross subject boundaries, so please check those multiple categories when searching for a specific toy.

Given the Masudaya reference material made available to the author, I have attempted to document the year of introduction of a toy if that information was provided. Some items appeared in Masudaya toy catalogs for multiple years. This information is included as observed. For very successful toys such as the locomotives and space explorers, this should be helpful in sorting out the years associated with the many variations of these toys.

Scarcity, Condition, and Values

Estimated current values for each toy are shown with the individual toy identification. Establishing value ranges is by no means an exact science due to the many variables and subjectivity involved. In the end, the market will determine the price realized at a sale, so the guide should be viewed as just that—a guide! The values published in this book represent a compilation of known sales at auction houses, from Internet auctions, and from sales at antique and specialized toy shows. In most cases they correspond to rarity of the toy and the market demand. The prices represent primarily North American values and should be considered in light of the following influences and the comments regarding interest and prices in other major geographic markets such as Japan and Europe.

Value Influences: Scarcity

Value is directly related to scarcity, with scarcity being determined by the *rarity* of the toy, *demand* for the toy and finally, the *condition* of the toy. Each toy listed contains a scarcity rating on a scale of 1 to 10 with 10 being the scarcest. Scarcity takes into account not only how many were produced but also how much of a demand exists among collectors. A rare toy with low production quantities can exist without much demand, just as a more plentiful toy can be in high demand and relatively more scarce

Value Influences: Rarity

Rarity is influenced primarily by how many toys were produced, how many survived, and in what part of the world they were marketed. To the manufacturer, the success of a toy was measured by sales. Poor selling toys were a financial disappointment to the manufacturer, but because of low supply and high demand, they result in higher prices for the collector today. Also, those individuals who had these toys as children, and their parents, did not realize they would become so collectible and valuable someday. Accordingly, toys were given away, thrown away, or often abandoned.

Toys move easily around the world today, but where the toys were marketed in earlier years impacts where they are today. The result is that toys made primarily for the North American market are harder to find in other parts of the world, causing the prices realized for those toys to be higher than that of North America. Conversely, toys made for the Japanese market are not often found in North America and toys made for the European market are also not as common in North America. This makes those toys more rare in North America, with the prices accordingly higher. And finally, some toys were only produced as samples that were taken by the trading companies to show importers what was available. Importers may have taken a sample but never placed an order, thus the toy never went into higher volume production.

Value Influences: Demand

The next major influence on values is demand. Some toys are just more popular than others based on the interests of collectors. Logic follows that toys sought by more collectors end up in more collections, making those toys harder to find.

Value Influences: Condition

Toys have survived in all stages of condition, from junk parts to near mint and in the original box. While the grading scale of C-1 to C-10 is widely utilized, there is certainly little consistency on how individuals grade their toys against this scale. If C-10 is truly mint (just as it came from the factory) and C-1 is parts, most collectors will search out toys in the C-6 to C-10 range. There is minimal interest among collectors for toys below C-6 (fine) condition. For vinyl and plastic toys, collectors usually prefer C-8 or higher. Accordingly, value estimates have been shown only for C-6 to C-10 toys, **all without boxes.**

C-6: *Fine*. Complete with no missing or broken parts. Nice condition with some evidence of aging and wear, but not played with hard.
C-7: *Very Fine*. Very minimal scratching and wear, but still bright.
C-8: *Excellent*. Very light general wear and appears close to new.
C-9: *Near Mint*. Looks like new, but upon close examination is not truly mint.
C-10: *Mint*. As it originally came from the factory with no defects, however, factory touch-up is acceptable.

20 *Scarcity, Condition, and Values*

Finding toys that are in factory mint condition is very difficult and these toys are uncommon! The best chance for this is when someone finds old unused store stock.

Value Influences: Boxes

Boxes were often discarded as the focus was on the toy. Also, because these toy boxes were not very strong, they were very susceptible to tearing and bending with handling. The toy protrusions often would poke a hole in the box. Moisture would cause the box staples to rust and children would write on the boxes. Still, the box art on many of the toys is very colorful and pictorial, making the boxes a significant item to the collector and very desirable in their own right. Sometimes boxes pictured the real item that the toy was modeled after, making for interesting graphics.

Boxes are *not* included in the value listing. However, boxes should be graded separately to accurately describe the condition of the toy and box and to better determine the approximate value of the toy and its box together. The C-1 to C-10 scale is appropriate for box grading.

Having the original box greatly enhances the value of the toy. Some rare C-10 boxes can easily double the value of a toy and some may add only 20% to the value. Be aware that reproduction boxes now exist for many toys and are generally worth maybe $10 to $25. Reputable dealers will identify when the box is a reproduction.

Aircraft

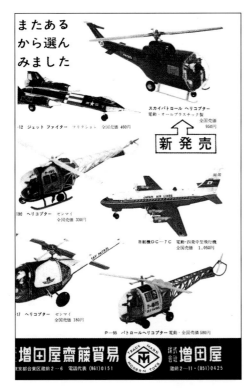

Trade journal ads, March 1968 (left) and July 1969.

Airplanes and other flying machines have always been a source of wonder and adventure, and were readily adapted as a subject for many toys. Children would make paper airplanes, planes from sheets of balsa wood, or just twigs, and pretend they could fly them. So it is no surprise that toy airplanes would be a popular subject for toy makers and an even more popular subject for the children who played with them. In this section, airplanes and helicopters are listed separately and airplanes are further separated by prop, jet, or military design.

Propeller Airplanes

#1732 **Loop Plane (Swallow)**. Swallow N-057 with remote controlled turnover action. R/C battery operated, tin. 8.5 in (22 cm). 1957-1958. Scarcity: 6. $100-$200.

22   *Aircraft*

#1784 **Loop The Loop Power Plane**. 7-inch Beechcraft A-104 flying on rotating battery control wire. Battery operated, tin. 22 in (56 cm). 1957-1958. Scarcity: 7. $150-$250.

#3253 **Condor Plane**. Condor 3253 with spinning prop and animal pilot. Friction, tin. 7.75 in (20 cm). 1962-1967. Scarcity: 5. $60-$100.

#1917 **Circling Plane N-108**. Condor N-108 with pilot, machine gun and spinning prop. Battery operated, tin. 8.5 in (22 cm). 1958. Scarcity: 6. $100-$200.

#3686 **Spark-Prop Plane Flying Tiger** with lighted prop, machine gun and vinyl pilot. Battery operated, tin with plastic. 9.5 in (24 cm). 1968-1971. Scarcity: 6. $125-$225.

#4177 **Looping Plane** with adjustable direction and turnover action. Battery operated, plastic. 9.5 in (24 cm). 1973-1975. Scarcity: 4. $40-$80.

Propeller Airplanes 23

#3925 **White Beechcraft Duke 60** with lithographed pilots, engine noise and turning propellers. Battery operated, tin with plastic. 13 in (33 cm). 1971-1972. Scarcity: 5. $100-$150.

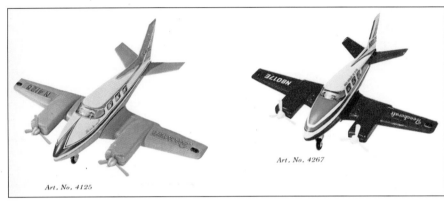

#3204 **Beechcraft Baron**. Beechcraft N-3204 with two props, plastic canopy and pilot. Friction, tin with plastic. 12 in (30 cm). 1962-1969. Scarcity: 4. $100-$175.

(left) #4125 **Beechcraft Plane Duke 60** with turning propellers. Friction, plastic with tin. 13 in (33 cm). 1973-1977. Scarcity: 3. $75-$150. (right) #4267 **Red Beechcraft Plane Duke 60** with lithographed pilots and passengers, turning propellers and adjustable direction. Battery operated, plastic with tin. 13 in (33 cm). 1974-1977. Scarcity: 3. $75-$150.

#3693 **Spark-Prop Beechcraft** with lighted props. Battery operated, tin with plastic. 12 in (30 cm). 1968-1973. Scarcity: 5. $125-$225.

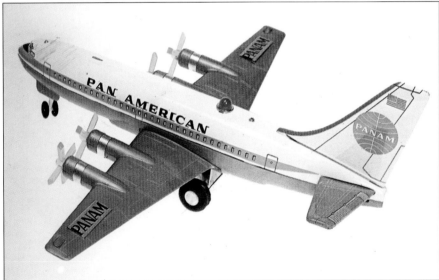

#3842 friction **Silver Beechcraft** with pilot and turning propellers or #3864 battery operated **Silver Beechcraft** with pilot, engine noise and turning propellers. Tin with plastic. 12 in (30 cm). 1971-1973. Scarcity: 4. $75-$125.

#3692 **Spark-Prop Pan American DC-7** with stop and go action, lighted and rotating props. Battery operated, tin with plastic. 17 in (43 cm). 1968-1971. Scarcity: 5. $175-$350.

24 *Aircraft*

#3701 **Pan American Douglas DC-7** with stop and go action and rotating props. Battery operated, tin with plastic. 17 in (43 cm). 1968-1973. Scarcity: 4. $175-$300.

Propeller Airplanes – Not Pictured

#3013 **Comic Plane.** "Lucky" plane 305 with monkey pilot. Friction, tin and plastic. 6.5 in (17 cm). 1960. Scarcity: 4. $60-$100.

#3201 **Planet Plane** with spinning prop, pilot and machine gun. R/C battery operated, tin. 9.5 in (24 cm). 1962-1965. Scarcity: 6. $100-$150.

#3289 **Comanche 180 Piper** with rotating prop. Friction, tin with plastic. 12.5 in (32 cm). 1964. Scarcity: 6. $100-$200.

#3311 **Silver Beechcraft** with rotating propellers and pilot in cockpit. Friction, tin with plastic. 12 in (30 cm). 1964-1968. Scarcity: 4. $100-$200.

#3400 **Beechcraft Plane** with rotating props and blinking light. R/C battery operated, tin with plastic. 12 in (30 cm). 1965-1969. Scarcity: 4. $100-$200.

#3654 **Japan Air Line Douglas DC-7** with stop and go action and spinning props. Battery operated, tin with plastic. 17 in (43 cm). 1968-1972. Scarcity: 5. $175-$300.

Jet Airplanes

#3719 **Spark-Prop Japan Air Lines DC-7** with stop and go action, lighted and rotating props. Battery operated, tin with plastic. 17.5 in (44 cm). 1969. Scarcity: 7. $200-$350.

#1496 **Sparkling Comet** with sparking. Friction, tin. 10 in (25 cm). 1955-1960. Scarcity: 5. $100-$175.

#4332 **Pan American Airplane w/Propellers.** DC-7 with four turning props. Friction, tin and plastic. 17.5 in (44 cm). 1975-1976. Scarcity: 5. $100-$200.

#3378 **Boeing 727 (large)** with stop and go action, sound and flashing lights. Battery operated, tin. 23 in (58 cm). 1965-1973. Scarcity: 4. $200-$325.

Jet Airplanes 25

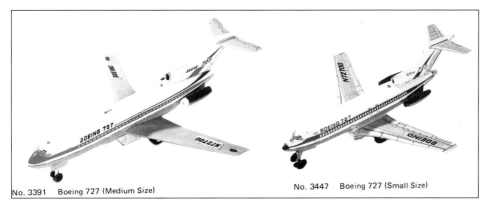

(left) #3391 **Boeing 727 (medium)** with sound and flashing light. Battery operated, tin. 19.75 in (50 cm). 1965-1973. Scarcity: 5. $175-$275.
(right) #3447 **Boeing 727 (small)** with sound and flashing light. Battery operated, tin. 13 in (33 cm). 1966-1973. Scarcity: 4. $125-$250.

#3571 **Radicon Boeing 727 Jet** with antenna and remote radio push button control. R/C battery operated, tin with plastic. 19.75 in (50 cm). 1968. Scarcity: 8. $200-$375.

(left) #3447 **Boeing 727 (small)** with sound and flashing light. Battery operated, tin. 13 in (33 cm). 1966-1973. Scarcity: 4. $125-$250. (center) #3448 **Boeing 737 (small)** with sound and flashing light. Battery operated, tin. 13 in (33 cm). 1966-1971. Scarcity: 5. $100-$200. (right) #4313 **Boeing 727 ANA (small)** with sound and flashing light. Battery operated, tin. 13 in (33 cm). 1960s. Scarcity: 7. $175-$300.

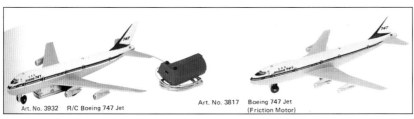

(left) #3932 **Jumbo Jet Boeing 747** with forward/reverse control. R/C battery operated, tin with plastic. 11 in (28 cm). 1971-1973. Scarcity: 4. $75-$175. (right) #3817 **Boeing 747 Jet**. Friction, tin and plastic. 11 in (28 cm). 1970-1977. Scarcity: 3. $75-$125.

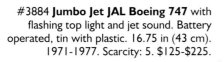

#3884 **Jumbo Jet JAL Boeing 747** with flashing top light and jet sound. Battery operated, tin with plastic. 16.75 in (43 cm). 1971-1977. Scarcity: 5. $125-$225.

26 Aircraft

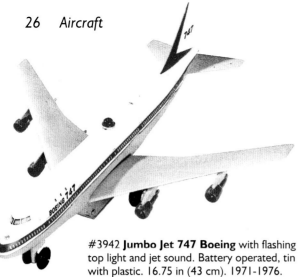

#3942 **Jumbo Jet 747 Boeing** with flashing top light and jet sound. Battery operated, tin with plastic. 16.75 in (43 cm). 1971-1976. Scarcity: 4. $100-$200.

(left) #3974 **Boeing 747 Jet - Pan American**. Friction, tin and plastic. 11 in (28 cm). 1971-1973. Scarcity: 6. $100-$150. (center) #3975 **Boeing 747 Jet - TWA**. Friction, tin and plastic. 11 in (28 cm). 1971-1973. Scarcity: 6. $100-$150. (right) #3976 **Boeing 747 Jet - United Airlines**. Friction, tin and plastic. 11 in (28 cm). 1971-1973. Scarcity: 6. $75-$125.

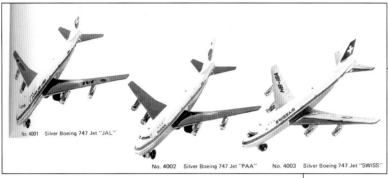

(left) #4001a **Silver Boeing 747 Jet - JAL** Japan Air Lines. Friction, tin with plastic. 11 in (28 cm). 1972-1973. Scarcity: 5. $75-$125. (center) #4002 **Silver Boeing 747 Jet - Pan Am**. Friction, tin with plastic. 11 in (28 cm). 1972-1973. Scarcity: 5. $75-$150. (right) #4003 **Silver Boeing 747 Jet - Swissair**. Friction, tin with plastic. 11 in (28 cm). 1973. Scarcity: 6. $100-$175.

#4152a **JAL Jumbo Jet 747** with sound and adjustable direction. Battery operated, plastic with tin. 26.75 in (68 cm). 1973. Scarcity: 7. $150-$250.

(left) #4043 **Boeing 747 Jet - JAL** with engine sound. Friction, tin and plastic. 16.75 in (43 cm). 1976-1977. Scarcity: 7. $150-$250. (right) #4001b **Boeing 747 Jet - JAL** Japan Air Lines with engine sound. Friction, tin and plastic. 11 in (28 cm). 1976. Scarcity: 6. $100-$200.

Jet Airplanes 27

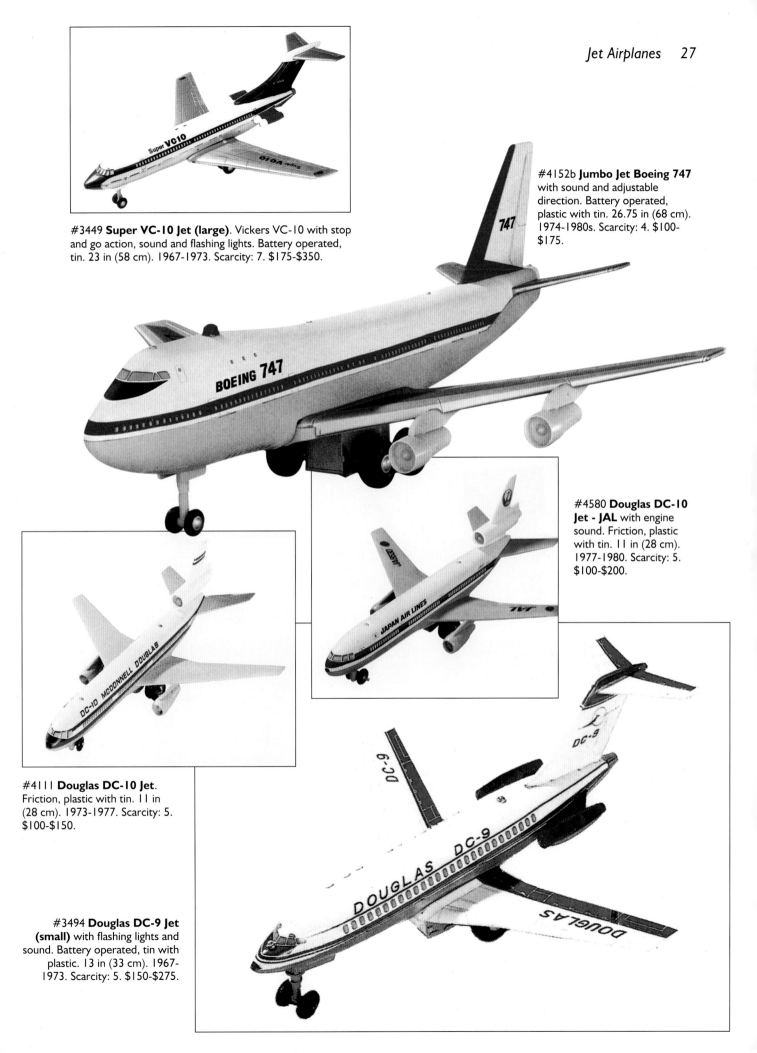

#3449 **Super VC-10 Jet (large)**. Vickers VC-10 with stop and go action, sound and flashing lights. Battery operated, tin. 23 in (58 cm). 1967-1973. Scarcity: 7. $175-$350.

#4152b **Jumbo Jet Boeing 747** with sound and adjustable direction. Battery operated, plastic with tin. 26.75 in (68 cm). 1974-1980s. Scarcity: 4. $100-$175.

#4580 **Douglas DC-10 Jet - JAL** with engine sound. Friction, plastic with tin. 11 in (28 cm). 1977-1980. Scarcity: 5. $100-$200.

#4111 **Douglas DC-10 Jet**. Friction, plastic with tin. 11 in (28 cm). 1973-1977. Scarcity: 5. $100-$150.

#3494 **Douglas DC-9 Jet (small)** with flashing lights and sound. Battery operated, tin with plastic. 13 in (33 cm). 1967-1973. Scarcity: 5. $150-$275.

28 Aircraft

Jet Airplanes – Not Pictured

#1521 **Comet 106** with sparking. Friction, tin. 8.5 in (22 cm). 1955-1960. Scarcity: 6. $75-$150.

#3385 **Boeing 727 (small)** with siren sound. Friction, tin. 13 in (33 cm). 1966-1971. Scarcity: 5. $150-$250.

#3424 **Boeing 737 (medium)** with flashing light. Battery operated, tin. 19 in (48 cm). 1966-1971. Scarcity: 5. $150-$250.

#3443 **Lufthansa Jet Boeing 727 (large)** with stop and go action, sound and flashing lights. Battery operated, tin. 23 in (58 cm). 1966-1970. Scarcity: 6. $200-$350.

#3472 **Douglas DC-9 Jet (medium)** with flashing lights and sound. Battery operated, tin with plastic. 19 in (48 cm). 1967-1971. Scarcity: 5. $175-$325.

#3475 **Lufthansa Jet Boeing 727 (small)** with flashing lights and sound. Battery operated, tin. 13 in (33 cm). 1966-1971. Scarcity: 6. $150-$275.

#3611 **Boeing 727 (medium)** with sound. Friction, tin. 19.75 in (50 cm). 1968-1971. Scarcity: 6. $150-$250.

Military Airplanes

#3309 **F-104 Missile Jet Fighter**. Lockheed F-104 with sparking and plastic missiles. Friction, tin with plastic. 13.75 in (35 cm). 1964-1968. Scarcity: 6. $125-$250.

#3267 **Jet Fighter "Silver Star."** Lockheed F-104 with sparking. Friction, tin. 13.5 in (34 cm). 1964-1968. Scarcity: 6. $125-$200.

#3655 **F-4 Phantom II Jet** with sound. Friction, tin. 14.5 in (37 cm). 1968-1973. Scarcity: 5d. $125-$200.

(left) #3233 **Lockheed Jet Fighter "Blue Star."** Lockheed F-104 with sparking. Friction, tin. 13.5 in (34 cm). 1962-1968. Scarcity: 5. $100-$200. (right) #3404 **Lockheed Jet Fighter Japan**. Lockheed F-104 with sparking. Friction, tin. 13.5 in (34 cm). 1965-1968. Scarcity: 6. $150-$250.

Military Airplanes 29

#3876 **Camouflage F-4 Phantom Jet - USAF** with jet sound. Friction, tin with plastic. 14.5 in (37 cm). 1971-1977. Scarcity: 3. $75-$125.

#3889 **Silver F-4 Phantom Jet - USAF** with flashing lights and jet sound. Battery operated, tin with plastic. 14.5 in (37 cm). 1971-1973. Scarcity: 5. $100-$175.

#3639 **YF-12 Jet Fighter**. Lockheed YF-12 experimental plane with sound. Friction, tin with plastic. 18.5 in (47 cm). 1968-1973. Scarcity: 5. $200-$300.

#3243 **Lockheed Hercules Giant Transport Plane (USAF)**. C-130 with spinning props and stop and go action, green color. Battery operated, tin. 16.5 in (42 cm). 1961-1967. Scarcity: 8. $800-$1,400.

Military Airplanes – Not Pictured

#1749 Jet Fighter F-100. USAF FW-549 jet fighter. Friction, tin. 6 in (15 cm). 1957-1958. Scarcity: 6. $125-$225.

#1771 U.S. Jet Fighter. USAF jet with rear propeller mechanism to propel the plane when hung from a hook. Windup, tin. 8 in (20 cm). 1957-1958. Scarcity: 6. $125-$225.

#3193 Lockheed Transport Plane. USAF C-130 silver color. Friction, tin. 16.5 in (42 cm). 1961. Scarcity: 9. $500-$900.

#3401 Missile Jet Fighter. Lockheed F-104 with sound and flashing light. R/C battery operated, tin. 13.75 in (35 cm). 1965-1968. Scarcity: 5. $100-$225.

Aero-Mini/Mini Air Series

Masudaya was the producer of a line of diecast airplanes known as *Aero Mini* during the early 1970s. These were produced under an agreement from an American company called Aero Mini Inc. The line of detailed scale model airplanes with liveries of airlines from around the world was very popular (now collectible) and appeared in the Masudaya Toy catalogs for four years. A similar *Mini Air* series appeared in the catalogs for another four years.

(left) **#3894 Aero Mini Boeing 737 ANA** - All Nippon Airways. Diecast. 5.5 in (14 cm). 1971. Scarcity: 9. $75-$125. (center) **#3895 Aero Mini Boeing 737 UAL** - United Airlines. Diecast. 5.5 in (14 cm). 1971-1973. Scarcity: 5. $40-$60. (right) **#3920 Aero Mini Boeing 737 Lufthansa** - German Airlines. Diecast. 5.5 in (14 cm). 1971-1972. Scarcity: 10. No price found.

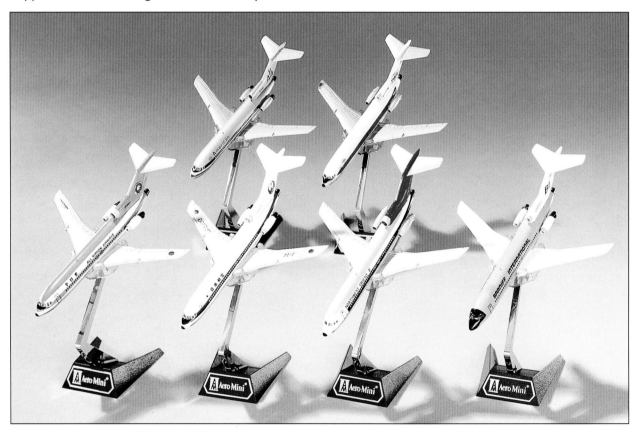

Front row (left to right):
#3892 Aero Mini Boeing 727 ANA - All Nippon Airways. Diecast. 6.75 in (17 cm). 1971-1973. Scarcity: 8. $75-$125.
#3890 Aero Mini Boeing 727 JAL - Japan Air Lines. Diecast. 6.75 in (17 cm). 1971-1973. Scarcity: 8. $75-$125.
#3893 Aero Mini Boeing 727 Northwest - Northwest Airlines. Diecast. 6.75 in (17 cm). 1971-1973. Scarcity: 5. $40-$60.

#3905 Aero Mini Boeing 727 Braniff - Braniff International. Diecast. 6.75 in (17 cm). 1971-1974. Blue color, + 25%. Scarcity: 5. $40-$60.
Back row (left to right):
#3953 Aero Mini Boeing 727 AA - American Airlines. Diecast. 6.75 in (17 cm). 1971-1974. Scarcity: 5. $40-$60.
#3891 Aero Mini Boeing 727 TWA - Trans World Airlines. Diecast. 6.75 in (17 cm). 1971-1974. Scarcity: 5. $40-$60.

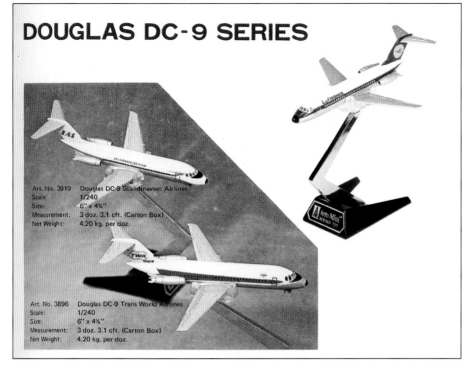

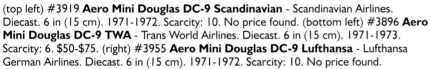

(top left) #3919 **Aero Mini Douglas DC-9 Scandinavian** - Scandinavian Airlines. Diecast. 6 in (15 cm). 1971-1972. Scarcity: 10. No price found. (bottom left) #3896 **Aero Mini Douglas DC-9 TWA** - Trans World Airlines. Diecast. 6 in (15 cm). 1971-1973. Scarcity: 6. $50-$75. (right) #3955 **Aero Mini Douglas DC-9 Lufthansa** - Lufthansa German Airlines. Diecast. 6 in (15 cm). 1971-1972. Scarcity: 10. No price found.

#3904 **Aero Mini Vickers VC-10 BOAC** - British Overseas Airways. Diecast. 8.5 in (22 cm). 1971-1973. Scarcity: 8. $75-$125.

(left) #3898 **Aero Mini Boeing 707 Northwest** - Northwest Airlines. Diecast. 7.5 in (19 cm). 1971-1972. Scarcity: 6. $50-$75. (right) #3897 **Aero Mini Boeing 707 Pan Am** - Pan American World Airways. Diecast. 7.5 in (19 cm). 1971-1973. Scarcity: 5. $45-$65.

(left) #3952/4004 **Aero Mini Boeing 707 AA** - American Airlines. Diecast. 7.5 in (19 cm). 1971-1974. Scarcity: 6. $50-$75. (right) #3906 **Aero Mini Boeing 707 TWA** - Trans World Airlines. Diecast. 7.5 in (19 cm). 1971-1974. Scarcity: 6. $50-$75.

(left) #3907 **Aero Mini Boeing 747 JAL** - Japan Air Lines. Diecast. 9.25 in (23 cm). 1971-1974. Scarcity: 8. $125-$175. (center) #3923 **Aero Mini Boeing 747 Alitalia** - Alitalia Line. Diecast. 9.25 in (23 cm). 1971. Scarcity: 8. $150-$200. (right) #3910 **Aero Mini Boeing 747 TWA** - Trans World Airlines. Diecast. 9.25 in (23 cm). 1971-1972. Scarcity: 7. $90-$150.

32 Aircraft

(left) **#3924 Aero Mini Boeing 747 Swissair** - Swissair Transport. Diecast. 9.25 in (23 cm). 1971. Scarcity: 10. No price found. (center) **#3908 Aero Mini Boeing 747 Pan Am** - Pan American World Airways. Diecast. 9.25 in (23 cm). 1971-1973. Scarcity: 6. $75-$125. (right) **#3909 Aero Mini Boeing 747 Northwest** - Northwest Airlines. Diecast. 9.25 in (23 cm). 1971-1972. Scarcity: 7. $90-$150.

#3954 Aero Mini Boeing 727 Lufthansa - Lufthansa German Airlines. Diecast. 6.75 in (17 cm). 1971. Scarcity: 10. No price found.

#3922 Aero Mini Boeing 747 Lufthansa - Lufthansa German Airlines. Diecast. 9.25 in (23 cm). 1971. Scarcity: 10. No price found.

(left to right):
#3996A Aero Mini McDonnell F-4E US Navy. Diecast. 7.5 in (19 cm). 1973-1974. Scarcity: 8. $100-$150.
#3996GB Aero Mini McDonnell F-4E British Royal Navy. Diecast. 7.5 in (19 cm). 1973-1974. Scarcity: 8. $100-$150.
#3996J Aero Mini McDonnell F-4E Japan. Diecast. 7.5 in (19 cm). 1972-1974. Scarcity: 8. $100-$150.
#3995J Aero Mini Lockheed F-104 Japan. Diecast. 5.25 in (13 cm). 1972-1974. Scarcity: 8. $100-$150.
#3995A Aero Mini Lockheed F-104 US Air Force. Diecast. 5.25 in (13 cm). 1973-1974. Scarcity: 8. $75-$125.
#3994 Aero Mini Mitsubishi Zero Fighter Japan. Produced in silver or green colors. Diecast. 5 in (13 cm). 1972-1974. Scarcity: 9. $150-$225.

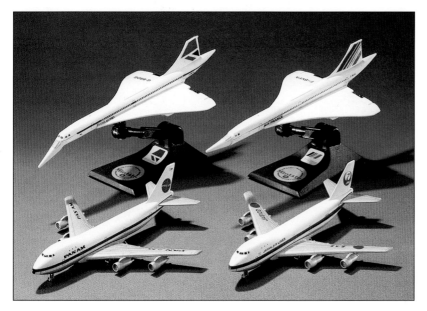

(back left) **#4601 Mini Air Concorde British Airways** - British Airways. Diecast. 7.5 in (19 cm). 1977-1979. Scarcity: 7. $75-$125. (back right) **#4556 Mini Air Concorde Air France** - Air France. Diecast. 7.5 in (19 cm). 1977-1979. Scarcity: 7. $75-$125. (front left) **#4555 Mini Air Boeing 747 PAA** - Pan American Airways. Diecast. 7.5 in (19 cm). 1977-1980. Scarcity: 7. $75-$125. (front right) **#4554 Mini Air Boeing 747 JAL** - Japan Air Lines. Diecast. 7.5 in (19 cm). 1977-1979. Scarcity: 8. $100-$150.

(left) **#4759 Mini Air Boeing 747 ANA** - All Nippon Airways. Diecast. 7.5 in (19 cm). 1980s. Scarcity: 8. $100-$150. (right) **#4705 Mini Air Lockheed Tri-Star ANA** - All Nippon Airways. Diecast. 7.5 in (19 cm). 1979-1980s. Scarcity: 8. $100-$150.

Aero-Mini – Not Pictured

#3921 Aero Mini Boeing 707 Lufthansa - German Airlines. Diecast. 7.5 in (19 cm). 1971-1972. Scarcity: 10. No price found.

#3989 Aero Mini Boeing 727 Pan Am - Pan American Airways. Diecast. 6.75 in (17 cm). 1972. Scarcity: 8. $60-$100.

#4117 Aero Mini Douglas DC-8 JAL - Japan Airlines. Diecast. 9 in (23 cm). 1974. Scarcity: 9. $125-$200.

Helicopters

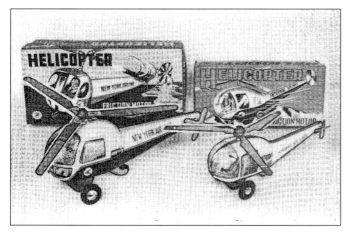

(left) **#1839 Helicopter-New York Airline** with spinning blades. Friction, tin. 10 in (25 cm). 1958-1960. Scarcity: 6. $100-$175. (right) **#1838 Helicopter-International Airline** with spinning blades. Friction, tin. 8.25 in (21 cm). 1958-1960. Scarcity: 5. $50-$75.

#1906 Helicopter HF-7 USAF H-60 with spinning rotor blades and pilot in cockpit. Friction, tin. 8 in (20 cm). 1958. Scarcity: 6. $75-$150.

34 Aircraft

#3059 **Twin Turbine Helicopter Flying Bus** with pilots, twin spinning blades and stop and go action. Battery operated, tin. 16.5 in (42 cm). 1960-1964. Scarcity: 7. $150-$300.

(left) #3224 **Rescue 24 Helicopter** whirls on ground with pilot and spinning rotors. Windup, tin. 9 in (23 cm). 1962-1967. Scarcity: 5. $50-$75. (right) #3617 **P-17 Sky Patrol Helicopter** whirls around with rotating propellers and pilot. Windup, tin with plastic. 9.5 in (24 cm). 1968-1973. Scarcity: 4. $50-$100.

#3326 **Army 26 Helicopter** whirls on ground with pilot and rotating propellers. Windup, tin with plastic. 10 in (25 cm). 1964-1966. Scarcity: 5. $85-$150.

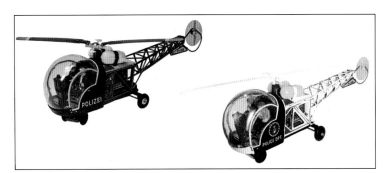

(left) #3499 **Bell-12 Helicopter "Polizei"** with two pilots, rotating blades and lights. Battery operated, mystery action, tin with plastic. 16.5 in (42 cm). 1967-1973. Scarcity: 5. $100-$200. (right) #3312 **Bell-12 Helicopter** with two pilots, rotating blades and lights. Battery operated, mystery action, tin with plastic. 16.5 in (42 cm). 1963-1977. Scarcity: 4. $100-$200.

#3351 **R-51 Rescue Helicopter**. Bell Rescue 51 whirls on ground with pilot and rotating blades. Windup, tin with plastic. 13.5 in (34 cm). 1965-1967. Scarcity: 6. $100-$175.

Helicopters 35

#3372 **Sikorsky Helicopter** with lithographed passengers in windows, sound and rotating propellers. Battery operated, mystery action, tin with plastic. 19 in (48 cm). 1965-1966. Scarcity: 8. $150-$275.

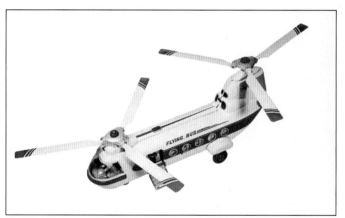

#3434 **Flying Bus Helicopter** with pilots, stop and go action and stewardess appearing at door. Battery operated, tin. 16.5 in (42 cm). 1966-1969. Scarcity: 8. $200-$350.

#3497 **Sky-Way Patrol Helicopter**. Bell traffic helicopter with pilot, rotating propellers and blinking lights. WHDH 850 in Boston. Battery operated, mystery action, tin with plastic. 16.5 in (42 cm). 1967-1969. Scarcity: 8. $150-$250.

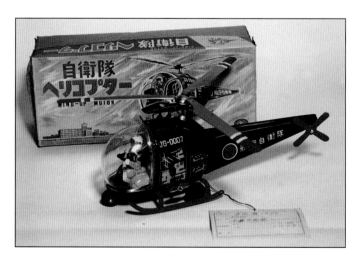

#3444 **Japanese Military Helicopter JG-0007** with pilot and clear cockpit. Windup, tin. 10 in (25 cm). 1966. Scarcity: 8. $125-$200.

#3573 **Army 73 Sikorsky Helicopter**. H-4820 Air Transport with soldiers in windows, sound and rotating propellers. Battery operated, mystery action, tin with plastic. 19 in (48 cm). 1967-1969. Scarcity: 6. $125-$225.

36 Aircraft

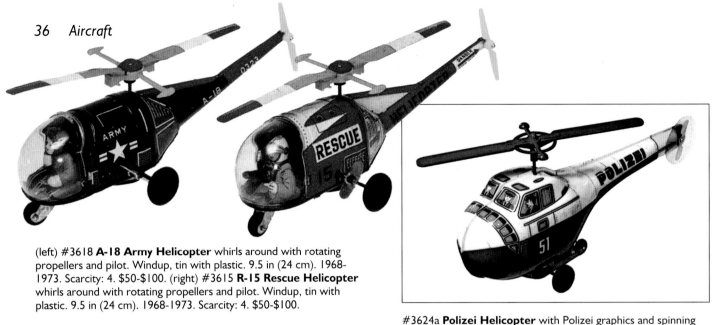

(left) #3618 **A-18 Army Helicopter** whirls around with rotating propellers and pilot. Windup, tin with plastic. 9.5 in (24 cm). 1968-1973. Scarcity: 4. $50-$100. (right) #3615 **R-15 Rescue Helicopter** whirls around with rotating propellers and pilot. Windup, tin with plastic. 9.5 in (24 cm). 1968-1973. Scarcity: 4. $50-$100.

#3624a **Polizei Helicopter** with Polizei graphics and spinning blades. Friction, tin with plastic. 9 in (23 cm). 1974-1975. Scarcity: 5. $50-$75.

#3619 **P-190 Patrol Helicopter** whirls around with rotating propellers, pilot and sound. Windup, tin with plastic. 14.5 in (37 cm). 1968-1970. Scarcity: 5. $100-$150.

#3710 **Traffic Control Helicopter** with lithographed controllers in windows, sound and rotating propellers. Battery operated, mystery action, tin with plastic. 19 in (48 cm). 1969. Scarcity: 9. $175-$325.

#3623 **Helicopters (6 assorted)**. Rescue, Fire Dept., Air Police, Air Mail, PAA, BOAC logos on 6 different helicopters with spinning rotor blades. Friction, tin with plastic. 9 in (23 cm). 1974-1979. Scarcity: 3. $30-$50. Also produced as windup versions (#3622), add 20%.

Helicopters 37

#3750 **Comic Helicopter** face shaped with eyes, nose and rotating blades. Windup, plastic. 4.75 in (12 cm). 1969-1973. Scarcity: 2. $10-$30.

#3917 **Hughes Helicopter** with revolving blades and lithographed interior images. Windup, plastic with tin. 9.25 in (23 cm). 1971-1973. Scarcity: 4. $30-$50.

#3812 **Hughes Helicopter** with pilot and passenger, stop and go action, rotating blades and automatic opening door. Battery operated, mystery action, plastic and tin. 16 in (41 cm). 1970-1972. Scarcity: 5. $50-$75.

#4010 **Police Patrol Helicopter** with revolving blades and engine noise. Windup, tin with plastic. 14.25 in (36 cm). 1975-1976. Scarcity: 5. $75-$150.

#4028 **Flying Bus Helicopter** with pilots, stop and go action, rotating blades and sound. Battery operated, tin with plastic. 16.5 in (42 cm). 1972-1974. Scarcity: 6. $100-$200.

38 Aircraft

(left) **#4631 Rescue Helicopter** with sound, whirling blade and flashing light in cockpit. Battery operated, mystery action, plastic with tin. 16 in (41 cm). 1978-1980s. Scarcity: 3. $40-$75. (right) **#4179 Hughes Helicopter** with sound, whirling blade and tin pilot. Battery operated, mystery action, plastic with tin. 16 in (41 cm). 1973-1980s. Scarcity: 3. $40-$80.

#4141 Gyrocopter with whirling blade. Windup, plastic with tin. 10.5 in (27 cm). 1973-1975. Scarcity: 4. $40-$60.

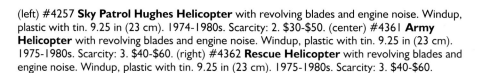

(left) **#4257 Sky Patrol Hughes Helicopter** with revolving blades and engine noise. Windup, plastic with tin. 9.25 in (23 cm). 1974-1980s. Scarcity: 2. $30-$50. (center) **#4361 Army Helicopter** with revolving blades and engine noise. Windup, plastic with tin. 9.25 in (23 cm). 1975-1980s. Scarcity: 3. $40-$60. (right) **#4362 Rescue Helicopter** with revolving blades and engine noise. Windup, plastic with tin. 9.25 in (23 cm). 1975-1980s. Scarcity: 3. $40-$60.

(left) **#4451 Mini Sky Helicopter** with rotating blades. Turns left and right via edges of key. Windup, plastic and tin. 5.5 in (14 cm). 1976-1980s. Scarcity: 3. $25-$45. (right) **#4474 Mini Police Patrol Helicopter** with rotating blades. Turns left and right via edges of key. Windup, plastic and tin. 5.5 in (14 cm). 1976-1979. Scarcity: 4. $30-$50.

Helicopters – Not Pictured

#1741 Helicopter No. 6. Rescue No.6 with pilot. Windup, tin. 10 in (25 cm). 1958. Scarcity: 5. $75-$150.

#1781A Rescue Helicopter No. 47. Rescue helicopter with pilot in cockpit and rotating blades. Windup, tin. 10 in (25 cm). 1957-1958. Scarcity: 5. $85-$150.

#1781B Navy HS 470 Helicopter with rotating blades and pilot in clear cockpit. Windup, tin. 10 in (25 cm). 1959-1964. Scarcity: 5. $100-$175.

#3142 Bell-47 Helicopter Police helicopter with pilot in cockpit and pontoons. Windup, tin and plastic. 13.5 in (34 cm). 1961-1964. Scarcity: 6. $100-$175.

#3182 Fire Patrol Helicopter No. 3182 with pilot and spinning rotors. Windup, tin with plastic. 8 in (20 cm). 1962. Scarcity: 6. $50-$75.

#3195 Patrol 95 Helicopter Bell 95 Patrol helicopter with spinning rotors, plastic skids, cockpit and pilot. Windup, tin and plastic. 13.5 in (34 cm). 1961-1964. Scarcity: 6. $100-$175.

#3656 Sky Patrol Helicopter with stop and go action, two pilots, blinking lights and rotating blades. Battery operated, plastic. 17.25 in (44 cm). 1969-1971. Scarcity: 3. $50-$75.

#3827 P-27 Sky Patrol Helicopter whirls around with pilot and revolving blades. Windup, tin with plastic. 14.5 in (37 cm). 1971-1976. Scarcity: 3. $50-$100.

Amusement Park & Kiddy Vehicle Toys

Every child and many adults loved going to the amusement park, or "Fairyland" as the Japanese often called it. The rides and make-believe associated with being at an amusement park also made an ideal subject for toys. The toys pictured in this section depict the fun associated with these parks. Also included in this section are play vehicles with "Kiddy" drivers representing the make-believe world of amusement park rides or backyard fun.

Amusement Park Rides

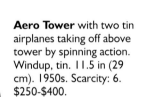

Aero Tower with two tin airplanes taking off above tower by spinning action. Windup, tin. 11.5 in (29 cm). 1950s. Scarcity: 6. $250-$400.

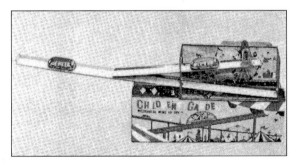

#1671 **Children Garden**. Over and under amusement park ride with two tin vehicles. Windup, tin. 17.25 in (44 cm). 1956-1960. Scarcity: 8. $175-$300.

#1933 **Fairy Land Train**. Animal locomotive No. 753 with clear plastic stack with bouncing balls. Friction, tin. 6.5 in (17 cm). 1959-1967. Scarcity: 1. $20-$30.

#1877 **Children Island** with two tin cars that go around amusement ride. Battery operated, tin. 19 in (48 cm). 1958. Scarcity: 10. $400-$800.

#3143 **Merry Land Hooting Train**. Steam train pulls passenger coach around platform base with station activating whistle. Battery operated, tin. 28 in (71 cm). 1961-1962. Scarcity: 7. $125-$200.

#3315 **Toyland Express/Fairy Train** with whistling sound and six tin figures that bounce around while train moves. Battery operated, mystery action, tin. 24.5 in (62 cm). 1964-1967. Also sold as Fairy Train. Scarcity: 5. $100-$200.

#3407 **Toyland Trolley**. Open top trolley with sound and six bouncing tin figures. Battery operated, mystery action, tin. 12 in (30 cm). 1965-1967. Scarcity: 6. $150-$275.

#3945 **Playland Coffee Cup**. Two figures in spinning cup with parasol. Battery operated, mystery action, plastic. 7.75 in (20 cm). 1971-1973. Scarcity: 5. $75-$125.

#3627 **Fairyland Choo Choo Loco** with two whistling trains and plane on wire circling on cylinder platform. Battery operated, tin. 9 in (23 cm). 1968-1969. Scarcity: 5. $75-$150.

Amusement Park Rides 41

#4139 **Playland** with rotating wheel that picks up cars and sets them on coaster track. Battery operated, plastic. 21.5 in (55 cm). 1973-1974. Scarcity: 5. $50-$80.

#4174 **Panda Bear Coffee Cup** with turning cup and parasol. Battery operated, mystery action, plastic. 9.5 in (24 cm). 1973. Scarcity: 6. $50-$100.

Amusement Park – Not Pictured

#1822 **Sight Seeing Plane** with two airplanes on crossbar flying around Eiffel Tower type structure. Battery operated, tin. 24 in (61 cm). 1958-1960. Scarcity: 10. $300-$600.

#1828 **Rope Way**. Twin cable cars travel back and forth to elevated station. Battery operated, tin. 9.5 in (24 cm). 1958-1960. Scarcity: 8. $250-$400.

#1929 **Jet Coaster**. Long car climbs and coasts over multi-level track. Windup, tin. 20 in (51 cm). 1959. Scarcity: 5. $100-$200.

#3031 **Fairy Land Bus**. Amusement park bus with face, driver and six wood-like kiddy figures. Battery operated, non-stop, tin. 10.5 in (27 cm). 1959-1962. Scarcity: 4. $75-$150.

#3080 **Fairy Land B-Z Fire Engine** with driver and two wood-like kiddy firemen. Battery operated, non-stop, tin. 10.5 in (27 cm). 1960-1962. Scarcity: 5. $100-$175.

#4159 **Train and Jet Playland** with flashing light, circling trains and rotating rocket rides. Battery operated, plastic. 9 in (23 cm). 1973. Scarcity: 7. $75-$125.

Kiddy Vehicles

#3929 **Sand Buggy (3 assorted colors)**. Comic buggy with driver and erratic bouncing and spinning action. Battery operated, plastic with tin. 10.75 in (27 cm). 1971-1978. Scarcity: 3. $40-$60.

#3927 **Prairie Comic Car**. Bouncing and spinning eccentric cars with driver. Windup, plastic with tin. 6 in (15 cm). 1971-1973. Scarcity: 4. $20-$40.

#4182 **Sand Buggy Sedan** with kiddy driver and eccentric car action. Battery operated, tin and plastic. 10.75 in (27 cm). 1973. Scarcity: 5. $40-$75.

#3933 **Puppy Pet Car** with vinyl head girl driver and squeaking puppies. Battery operated, mystery action, tin with vinyl. 10.5 in (27 cm). 1971-1972. Scarcity: 6. $75-$125.

#3168 **Non Fall Fire Chief Car** with oversized fireman with vinyl head and bell. Battery operated, non-fall, tin. 9 in (23 cm). 1961-1964. Scarcity: 6. $100-$200.

#3737 **Police Hand Car** with police boy rowing and bell sound. Battery operated, mystery action, tin with vinyl. 10 in (25 cm). 1969-1973. Scarcity: 4. $125-$200.

Kiddy Vehicles 43

(left) **#3798 Baby Police Car** with oversized baby driver and bell sound. Windup, plastic with tin. 5 in (13 cm). 1970-1973. Scarcity: 2. $20-$40. (right) **#3802 Baby Fire Engine** with oversized baby driver and bell sound. Windup, plastic with tin. 5.25 in (13 cm). 1970-1971. Scarcity: 3. $20-$40.

#3809 **Siren Patrol Car**. Mercedes 230 SL with moving kiddy vinyl head driver, trunk siren and hood light. Battery operated, mystery action, tin with vinyl. 13 in (33 cm). 1970-1973. Scarcity: 3. $100-$150.

#3811 **Siren Fire Car**. Mercedes 230 SL with moving kiddy vinyl head driver, trunk siren and hood light. Battery operated, mystery action, tin with vinyl. 13 in (33 cm). 1970-1972. Scarcity: 4. $125-$175.

#3830 **Flag Fire Engine** with oversize vinyl fire driver that moves flag and turns head while bell rings. Plastic siren and flashing light. Battery operated, mystery action, tin and vinyl. 10.5 in (27 cm). 1970-1973. Scarcity: 4. $75-$125.

#3839 **Baton Police Car** with oversize vinyl police driver that moves baton and turns head. Crank siren. Battery operated, mystery action, tin. 10.5 in (27 cm). 1970-1973. Scarcity: 4. $75-$125.

(left) #3940 **Police Patrol Car with Steering** with vinyl head kiddy police driver that steers and large trunk mounted siren. Battery operated, mystery action, tin. 10.5 in (27 cm). 1971-1972. Scarcity: 5. $75-$125. (right) #3941 **Fire Engine Car with Steering** with vinyl head kiddy fire driver and bell. Battery operated, mystery action, tin. 10.5 in (27 cm). 1971-1973. Scarcity: 4. $75-$125.

44 Amusement Park & Kiddy Vehicle Toys

#4138 **Kiddy Police Car** with kiddy police driver, sound and flashing light. Battery operated, mystery action, tin and plastic. 10.5 in (27 cm). 1973-1980s. Scarcity: 2. $45-$85.

#3158 **Go Kart**. Vinyl head tin boy riding in Go Kart. Friction, tin with vinyl. 6 in (15 cm). 1962-1964. Scarcity: 4. $75-$150.

(left) #4268 **Kiddy Fire Chief Car** with kiddy fire driver, sound and flashing light. Battery operated, mystery action, tin and plastic. 10.5 in (27 cm). 1974-1976. Scarcity: 4. $50-$100. (right) Cat. #4138. **Kiddy Police Car**.

#3173 **Go Kart**. Vinyl head tin boy in Go-Kart with mechanical remote control wire. Battery operated, tin with vinyl. 6.5 in (17 cm). 1962. Scarcity: 6. $125-$225.

#3115 **Hot-Rod** with siren and tin kiddy driver with vinyl head. Friction, tin. 6 in (15 cm). 1961-1962. Scarcity: 5. $100-$175.

#3212 **Go Kart**. Boy driver with vinyl head rides in Go-Kart that starts and stops by pulling string connected to switch. Battery operated, tin with vinyl. 6.5 in (17 cm). 1962. Scarcity: 6. $100-$200.

Kiddy Vehicles 45

#3296/4087 **Hand Car** with tin boy driver and ringing bell. Battery operated, mystery action, tin with vinyl. 8 in (20 cm). 1963-1973. Renumbered in 1973 to #4087. Scarcity: 4. $150-$225.

#4202 **Piston Power Go Kart** with driver and lighted moving pistons. Battery operated, mystery action, plastic. 9 in (23 cm). 1973-1976. Scarcity: 3. $60-$100.

#3987 **Champion Race Car** with kiddy driver and engine sound. Friction, tin and plastic. 12 in (30 cm). 1972. Scarcity: 5. $50-$100.

#4348 **Hand Car** with vinyl head boy moving in rowing manner with ringing bell. Battery operated, mystery action, tin with plastic. 8 in (20 cm). 1975-1976. Scarcity: 5. $175-$250.

#4201 **Kiddy Race Car** with kiddy race driver, sound and flashing light. Battery operated, mystery action, tin and plastic. 10.5 in (27 cm). 1973. Scarcity: 6. $60-$100.

#3183 **Bob Sleigh** with vinyl head tin boy on sled with propeller. Looks like an airboat. Friction, tin with vinyl. 6.25 in (16 cm). 1962-1964. Scarcity: 4. $100-$175.

46 *Amusement Park & Kiddy Vehicle Toys*

#4279 **Kiddy Race Car** with kiddy race driver, sound and flashing, visible cylinder. Battery operated, mystery action, tin and plastic. 10.5 in (27 cm). 1974-1978. Scarcity: 3. $50-$75.

#4812 **Kiddy Piston Race Car** with moving piston rods and engine sound. Battery operated, plastic. 9 in (23 cm). 1980-1980s. Scarcity: 3. $30-$50.

#3681 **Overland Special Loco** with kiddy engineer that turns head and swings lantern. Loud whistle. Battery operated, mystery action, tin and plastic. 10 in (25 cm). 1968-1974. Scarcity: 3. $50-$75.

#2875 **Baby Racer** with oversized boy driver and bell sound. Windup, plastic with tin. 7 in (18 cm). 1970-1972. Scarcity: 4. $20-$30.

#3793 **Mini Overland Express** with vinyl driver and lantern and shaking smoke stack. Battery operated, mystery action, plastic with tin. 6.5 in (17 cm). 1970-1971. Scarcity: 4. $20-$40.

Kiddy Vehicles 47

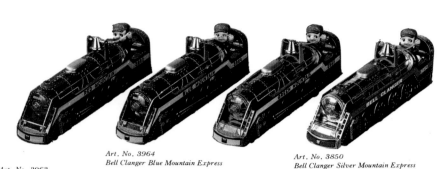

Art. No. 3963 **Bell Clanger Red Mountain Express**
Art. No. 3964 **Bell Clanger Blue Mountain Express**
Art. No. 3965 **Bell Clanger Green Mountain Express**
Art. No. 3850 **Bell Clanger Silver Mountain Express**

(left to right) #3963 **Bell Clanger Red Mountain Express** with boy vinyl driver pulling bell. Battery operated, mystery action, tin with plastic. 15.5 in (39 cm). 1971-1980s. Scarcity: 1. $20-$40. #3964 **Bell Clanger Blue Mountain Express** with boy vinyl driver pulling bell. Battery operated, mystery action, tin with plastic. 15.5 in (39 cm). 1971-1980s. Scarcity: 1. $20-$40. #3965 **Bell Clanger Green Mountain Express** with boy vinyl driver pulling bell. Battery operated, mystery action, tin with plastic. 15.5 in (39 cm). 1971-1980s. Scarcity: 1. $20-$40. #3850 **Bell Clanger Silver Mountain Express** with boy vinyl driver pulling bell. Battery operated, mystery action, tin with plastic. 15.5 in (39 cm). 1970-1980s. Scarcity: 1. $20-$40.

#3868 **Engineer Locomotive**. Comic faced engine with boy swinging lantern, sound and whistle. Battery operated, mystery action, tin with plastic. 9.5 in (24 cm). 1971-1975. Scarcity: 2. $30-$50.

#3877 **Flag Bell Ringer** with vinyl boy head, train sound, whistle and arm waving flag. Battery operated, mystery action, tin with plastic. 10 in (25 cm). 1971-1973. Scarcity: 3. $40-$60.

#3878 **Flag Western Locomotive** with vinyl boy head, train sound, whistle and arm waving flag. Battery operated, mystery action, tin with plastic. 11 in (28 cm). 1971-1973. Scarcity: 3. $50-$75.

#3912 **Cannonball Express** with vinyl boy head, train sound, headlight, smoke and arm waving flag. Battery operated, mystery action, tin with plastic. 11 in (28 cm). 1971-1973. Scarcity: 3. $50-$75.

#3930 **Fairyland Loco** with rocking tin boy on top of engine, whistle, lighted funnel and train noise. Battery operated, mystery action, tin with plastic. 13 in (33 cm). 1971-1973. Scarcity: 5. $75-$125.

48 Amusement Park & Kiddy Vehicle Toys

#4008 **Kiddy Locomotive** with kiddy engineer, moving smokestack, steering and forward/reverse action. R/C battery operated, plastic with tin. 6.5 in (17 cm). 1972-1974. Scarcity: 3. $20-$40.

#4319 **Fireball Express** with kiddy driver waving lantern. Battery operated, mystery action, tin with plastic. 15.5 in (39 cm). 1975. Made for Sonsco. Scarcity: 5. $50-$75.

#4044 **Dolly Super Express** with whistle sound, headlights and kiddy engineer moving head and hand. Battery operated, mystery action, tin with plastic. 11 in (28 cm). 1972-1974. Scarcity: 4. $50-$75.

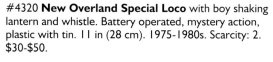

#4320 **New Overland Special Loco** with boy shaking lantern and whistle. Battery operated, mystery action, plastic with tin. 11 in (28 cm). 1975-1980s. Scarcity: 2. $30-$50.

#4337 **New Dolly Super Express**. Bullet train with whistle sound, headlights and boy engineer moving hand with flag. Battery operated, mystery action, tin with plastic. 11 in (28 cm). 1975-1977. Scarcity: 4. $50-$75.

Kiddy Vehicles 49

#4450 **Mini Engineer Locomotive** with boy shaking lantern, whistle sound and flashing light in stack. Battery operated, mystery action, plastic with tin. 6.25 in (16 cm). 1976-1980s. Scarcity: 2. $20-$30.

#3848 **Happy Kiddy Trolley** with driver that moves body and rings bell. Battery operated, mystery action, plastic with tin. 11 in (28 cm). 1970-1974. Scarcity: 3. $40-$60.

#4276 **Bell Express Red** with moving boy pulling bell chain, whistling sound and flashing headlight. Battery operated, mystery action, tin and plastic. 11.5 in (29 cm). 1974-1975. Scarcity: 3. $30-$50.

#2962 **Tiny Dump Truck** with oversized boy driver and bell sound. Windup, plastic with tin. 7 in (18 cm). 1970-1972. Scarcity: 4. $20-$30.

Kiddy Vehicles – Not Pictured

#2178 **Playland Police and Fire Chief** with mystery action. Windup, tin and plastic. 4.5 in (11 cm). 1979-1980. Scarcity: 4. $20-$30.

#3771 **Fireball Express** with whistle and bell sound, kiddy engineer and blinking lantern. Battery operated, mystery action, tin with plastic. 16 in (41 cm). 1969-1972. Scarcity: 2. $30-$50.

#4203 **New Sound Police Car** with kiddy police driver, antenna, sound and light. Battery operated, mystery action, tin with plastic. 10.5 in (27 cm). 1973. Scarcity: 6. $100-$150.

#4204 **New Fire Engine Car** with kiddy fire driver, bell, sound and light. Battery operated, mystery action, tin with plastic. 10.5 in (27 cm). 1973. Scarcity: 6. $100-$150.

#4277 **Bell Express Blue** with moving boy pulling bell chain, whistling sound and flashing headlight. Battery operated, mystery action, tin and plastic. 11.5 in (29 cm). 1974-1975. Scarcity: 3. $30-$50.

#4278 **Bell Express Silver** with moving boy pulling bell chain, whistling sound and flashing headlight. Battery operated, mystery action, tin and plastic. 11.5 in (29 cm). 1974-1979. Scarcity: 3. $30-$50.

Animals

Animals have been the subjects of toys for as long as people can remember. Over time, the toys became more lifelike and animated. In 1949, the *Walking Bear* was introduced (Item No.1112). It was made of a pressed tinplate body covered with plush and powered by a windup clockwork spring mechanism. It walked with realistic motion and was very cute. Masudaya believes it to be one of the first animated and articulated animal toys. This was during the time that the toy industry resumed operations after the desperation of World War II, so conditions were very poor. Materials were in short supply and some of the bodies were made from beverage cans opened and flattened with a hammer.

Fortunately, the product became a huge hit, making this good seller a factor in Masudaya's recovery during this rebuilding period. The successful *Walking Bear* was followed by a *Walking Elephant* (Item #1232) and a walking *Jolly Penguin* (Item #1308), both of which became good sellers as well. Since then, many animals have been made both with windup and battery operated mechanisms. Today, animals continue to be one of the important categories in toy stores.

Animals are shown here by subject, with the most popular animal toy subjects in separate listings and the balance shown under "Other Animals."

Trade journal ad, October 1966.

Bears

College Bear with mortar board cap, tie and cane. Windup, plush over tin. 6 in (15 cm). 1950s. Scarcity: 5. $40-$75.

#1112 **Walking Lovely Bear**. Chenille covered bear that walks on all four legs. Windup, plush over tin. 6 in (15 cm). 1949. Also sold as Clever Bear, Roaring Bear, etc. Scarcity: 2. $40-$80.

Bears 51

Grand-Pa Panda Bear. Panda bear with lighted eyes eating popcorn. Battery operated, tin and plush over tin. 10 in (25 cm). 1960. Scarcity: 6. $225-$425.

#1863 **Bear Target Game** with plastic dart pistol and two darts. Shoot his tummy and he plays the drum.. Battery operated, tin. 9 in (23 cm). 1958. Scarcity: 6. *Courtesy of Smith House Toys.* $200-$350.

#1919 **Bubble Bear** with bubble blowing pipe. Battery operated, tin. 9.5 in (24 cm). 1958-1960. Scarcity: 6. $200-$400.

#1925 **Father Bear** reading book and drinking in rocking chair. Designed to match knitting Mother Bear. Battery operated, tin and plush. 10 in (25 cm). 1959. Scarcity: 6. $175-$350.

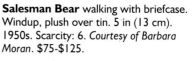

Salesman Bear walking with briefcase. Windup, plush over tin. 5 in (13 cm). 1950s. Scarcity: 6. *Courtesy of Barbara Moran.* $75-$125.

52 Animals

#3021 **Telephone Bear**. Bear on chair with old time telephone. Battery operated, plush and tin. 9.5 in (24 cm). 1960. Scarcity: 8. $200-$400.

#3040 **My Little Bear** with Bottle, Book, Bowl or Candy. Windup, plush over tin. 5.25 in (13 cm). 1959-1962. Scarcity: 3. $20-$40.

#3647 **Walking Bear** walks then stops and roars. R/C battery operated, plush. 13 in (33 cm). 1968-1973. Scarcity: 3. $30-$50.

#4165 **Panda Baby Car** with ringing bell. Windup, plastic with tin. 7 in (18 cm). 1973-1974. Scarcity: 4. $30-$60.

#4126 **Happy Panda Bear Family**. Balloon waves and bears shake their heads with sound. Battery operated, mystery action, plush and plastic. 12 in (30 cm). 1973-1978. Scarcity: 3. $40-$80.

#3151 **Bear the Cashier** with bear working adding machine and answering phone. Battery operated, plush and tin. 7.5 in (19 cm). 1961-1962. Scarcity: 7. $200-$400.

#4205 **Baby Panda Tricycle** with ringing bell. Windup, plastic with tin. 6 in (15 cm). 1974. Scarcity: 4. $30-$60.

#4594 **Walking Bear**. Bear with newspaper walks while ringing bell. Battery operated, plush. 10.5 in (27 cm). 1977-1979. Scarcity: 2. $25-$50.

#4226 **Bear Carriage** with two bears in chairs that revolve with umbrella when pulled. Pull toy, plastic. 14.25 in (36 cm). 1974-1977. Scarcity: 2. $20-$40.

#4689 **Guitarist Bear** playing musical guitar. Battery operated, plush with plastic. 13 in (33 cm). 1979-1980. Scarcity: 4. $30-$60.

#4621 **Walking Bear**. Bear with sandwich walks while ringing bell. Battery operated, plush. 10.5 in (27 cm). 1978-1980. Scarcity: 2. $25-$50.

#4654 **Tricycler Panda Bear** with vinyl panda family and ringing bell. Windup, vinyl and plastic. 7 in (18 cm). 1978-1980s. Scarcity: 3. $15-$30.

54 Animals

#4711 **Holiday Panda** with baby panda on rocker. Battery operated, plush and plastic. 9 in (23 cm). 1979-1980s. Scarcity: 2. $25-$45.

#4735 **Drummer Panda Bear** with moving face and playing drums. Battery operated, plush with plastic. 12 in (30 cm). 1979-1980s. Scarcity: 4. $30-$60.

#7321 **Jumping Koala Bear** hops forward. Windup, plush. 7.5 in (19 cm). 1972-1978. Scarcity: 3. $20-$40.

Bears – Not Pictured

Giant Roaring Bear. Chenille covered bear that walks on all four legs with sound. Windup, plush over tin. 6 in (15 cm). 1950. Scarcity: 2. $40-$80.

Make-up Bear. Bear in rocking chair with mirror make-up. Battery operated, plush and tin. 9 in (23 cm). 1959. Scarcity: 8. $350-$700.

Skating Bear. Plush bear on tin roller skates. Windup, tin with plush. 7 in (18 cm). 1940s. Scarcity: 5. $150-$300.

#1921 **Roaring Walking Bear** walks with roaring sound. R/C battery operated, plush over tin. 8.5 in (22 cm). 1958. Scarcity: 5. $50-$100.

#1923 **Mother Bear (Knitting Bear)** knitting on rocking chair. Designed to match reading Father Bear. Battery operated, plush and tin. 10 in (25 cm). 1959. Scarcity: 4. $125-$250.

#1959 **Ice-cream Baby Bear** eating chocolate ice cream in chair (vanilla ice cream add 20%). Battery operated, plush and tin. 9 in (23 cm). 1960. Scarcity: 8. $250-$500.

#1972 **Climbing Bear Target Game** with dart gun to shoot at bear climbing tree. Tin and cardboard. 14 in (36 cm). 1960. Scarcity: 7. $100-$150.

#3026 **Reading Bear** sitting with cup in one hand and book in the other. Windup, tin and plush. 8 in (20 cm). 1960. Scarcity: 4. $75-$125.

#3027 **Knitting Bear**. Sitting mother bear knitting. Windup, tin and plush. 8 in (20 cm). 1960. Scarcity: 4. $75-$125.

#4784 **Drummer Bear** with shaking body while playing drum. Battery operated, plush and plastic. 15 in (38 cm). 1980-1980s. Scarcity: 3. $40-$75.

#3028/1 **Knitting Bear On A Stool** Mother bear knitting while sitting on stool. Windup, tin and plush. 8.5 in (22 cm). 1960. Scarcity: 5. $75-$150.

#3028/2 **Reading Bear On A Stool**. Sitting on stool with cup in one hand and book in the other. Windup, tin and plush. 8.5 in (22 cm). 1960. Scarcity: 5. $75-$150.

#3078 **Jolly Bear with Robin**. Bear sits with robin on hand. Battery operated, plush and tin. 10 in (25 cm). 1960-1962. Scarcity: 9. $300-$600.

#3119 **Bear on Scooter**. Plush bear on flat 3-wheel scooter. Windup, tin and plush. 6 in (15 cm). 1961-1962. Scarcity: 2. $25-$50.

#3132 **Lite-up Telephone Bear**. Bear on seat with lite-up modern telephone. Battery operated, plush and tin. 10 in (25 cm). 1960-1962. Scarcity: 7. $150-$300.

#3136 **Bear on Tree**. Plush bear swings on wire stand in a variety of positions. Windup, plush and tin. 7 in (18 cm). 1961-1964. Scarcity: 5. $75-$125.

#3394 **Lovely Bear** with fish in its mouth and walking action. Windup, plastic and tin. 5.25 in (13 cm). 1965-1971. Scarcity: 2. $20-$30.

#3527 **Pretty Bear** walking with umbrella. Windup, plastic. 7.5 in (19 cm). 1967-1969. Scarcity: 2. $20-$30.

#4166 **Panda Bear Cyclist** with ringing bell. Windup, plastic with tin. 7 in (18 cm). 1973-1974. Scarcity: 4. $30-$60.

#4325 **Happy Family Bear**. Bear pulling cart with cubs. Moving heads and balloon along with yelping sounds. Battery operated, mystery action, plush with plastic. 12 in (30 cm). 1975-1977. Scarcity: 4. $50-$100.

#4736 **New Family Panda Bear** with bear cubs in cart, moves mouth and arm and cubs cry. Battery operated, mystery action, plush and plastic. 12 in (30 cm). 1979-1980. Scarcity: 4. $50-$100.

#7159 **Jumping Bear** hops forward. Windup, plush. 8 in (20 cm). 1972-1978. Scarcity: 2. $20-$40.

#7394 **My Lovely Bear** with moving head. Windup, vinyl. 5.5 in (14 cm). 1974-1975. Scarcity: 2. $20-$30.

Birds

#2590 **Musical Bird in Cage with Jewelry Box** has jewelry box, dancing bird, musical movement, and rotating cage. Windup, tin. 11 in (28 cm). 1969-1972. Scarcity: 4. $35-$75.

#7030 **Musical Bird in Cage** with dancing bird, musical movement and rotating cage. Bird changes color with change in temperature. Windup, tin. 8 in (20 cm). 1970-1975. Scarcity: 3. $35-$75.

#7032 **Musical Bird in Cage** with dancing bird, musical movement and rotating cage. Windup, tin. 8 in (20 cm). 1970-1975. Scarcity: 3. $35-$75.

#3379 **Happy Singing Bird** with bird in cage singing and moving on branch. Battery operated, tin and celluloid. 9.5 in (24 cm). 1965-1969. Masudaya continued to produce various bird in cage toys into the 1980s. Scarcity: 3. $75-$125.

#7031 **Musical Bird in Cage** with dancing bird, musical movement and rotating round cage. Bird changes color with change in temperature. Windup, tin. 8 in (20 cm). 1970-1975. Scarcity: 3. $35-$75.

#7098 **Bird in Cage with Music** with chirping and dancing bird. Windup, tin. 8.5 in (22 cm). 1971-1973. Scarcity: 4. $35-$75.

56   Animals

#7033 **Musical Bird in Cage with Jewelry Box** with pull out drawer music box, dancing bird, musical movement and rotating cage. Windup, tin. 8 in (20 cm). 1970-1973. Scarcity: 3. $35-$75.

Singing Chicken clucks while pecking at seed. Windup, tin. 3 in (8 cm). 1940s. Scarcity: 5. $75-$125.

#3423 **Tiny Chickling** with walking action. Windup, plush and tin. 5.75 in (15 cm). 1966-1973. Scarcity: 2. $20-$30.

#3396/3776 **Family Duck** with three babies following and connected by chain. Windup, plastic with tin. 12.5 in (32 cm). 1965-1978. Scarcity: 3. $30-$50.

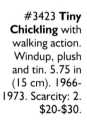

#3435 **Quacking Duck** with quacking sound. Battery operated, non-fall, tin. 9 in (23 cm). 1966-1967. Scarcity: 4. $40-$80.

#3800 **Quacking Duck**. Walking duck with quacking sound. Windup, plush and tin. 6 in (15 cm). 1970-1972. Scarcity: 3. $25-$65.

Birds 57

#4014 **Strolling Duck** with opening mouth. Windup, plastic with tin. 4 in (10 cm). 1972-1976. Scarcity: 2. $20-$40.

#4349 **Pull Toy Duck** on wheels with crying voice. Pull toy, plush. 13.5 in (34 cm). 1976. Scarcity: 4. $15-$30.

#4019 **Family Duck**. Mother duck opens mouth and quacks while pulling two baby ducks. Windup, plush and plastic. 5.75 in (15 cm). 1972-1974. Scarcity: 3. $35-$75.

#3520 **Jolly Penguin and Baby**. Walking penguin with baby following. Windup, plush and plastic. 5.25 in (13 cm). 1967. Scarcity: 5. $40-$75.

#4057 **Strolling Duck Family**. Mother duck opens mouth and quacks with baby duck moving on her back. Battery operated, mystery action, plush. 7.75 in (20 cm). 1972-1976. Scarcity: 2. $25-$45.

#1308 **Jolly Penguin** walks with turning head and flapping wings. Windup, plush with tin. 5.25 in (13 cm). 1951-1960. Scarcity: 3. $75-$125.

Birds – Not Pictured

Wobbling Pelican. Pelican on eccentric wheels. Windup, celluloid. 4 in (10 cm). 1940s. Scarcity: 4. $75-$150.

#1045 Strolling Duck. Duck on eccentric wheels creating wobbly effect. Windup, celluloid. 5 in (13 cm). 1940s. Scarcity: 4. $75-$150.

#1347 Quacking Duck that walks and quacks. Windup, plush over tin. 7 in (18 cm). 1952-1959. Also see #1733 (next entry). Scarcity: 3. $40-$80.

#1733 Quack Quack Duck that walks and quacks. Windup, plush over tin. 7 in (18 cm). 1957-1958. Reissue of #1347. Scarcity: 2. $40-$80.

#1792 Duck. Walking duck with jacket, tie and cap. R/C battery operated, tin with plush. 8 in (20 cm). 1957-1958. Scarcity: 3. $30-$60.

#1812 Jolly Penguin with Lighted Eyes. Penguin walks with lighted eyes and flapping wings. Battery operated, tin and plush. 7.5 in (19 cm). 1957-1961. Scarcity: 3. $50-$100.

#1819 Happy Singing Bird with bird in cage singing and moving on branch. Battery operated, tin and celluloid. 9 in (23 cm). 1957-1961. Reissued as #3379. Scarcity: 4. $75-$125.

#3038 Jumping Pet. Chicken, rabbit, squirrel, or dog hop on their feet. Windup, tin. 3 in (8 cm). 1959-1964. Scarcity: 3. $20-$40.

#3099 Happy Time Duckling. Mother duck with duckling on her back. Windup, plush over tin. 7.5 in (19 cm). 1960-1964. Scarcity: 2. $40-$75.

#3139 Happy Duck. Quacking duck with moving bill. Windup, plush over tin. 5.5 in (14 cm). 1961-1962. Scarcity: 2. $25-$50.

#3248 Jolly Penguin. Walking penguin with blinking eyes and screech sound. R/C battery operated, plush and tin. 8.75 in (22 cm). 1964-1966. Scarcity: 5. $40-$80.

#3249 Parrot. Walking parrot with blinking eyes and screech sound. R/C battery operated, plush and tin. 8.75 in (22 cm). 1964-1968. Scarcity: 5. $50-$100.

#3361 Lovely Duckling with walking action. Windup, plastic and tin. 5.5 in (14 cm). 1965-1974. Scarcity: 2. $20-$40.

#3514 Merry Duck walks with moving beak. Windup, plastic with tin bill. 5.25 in (13 cm). 1968-1971. Scarcity: 2. $20-$40.

#3532 Tiny Duckling with walking action. Windup, plush with tin. 5.25 in (13 cm). 1967-1973. Scarcity: 2. $15-$25.

#3585 Strolling Duck (two styles) walks with moving mouth. Windup, plastic with tin. 5.25 in (13 cm). 1968-1971. Scarcity: 2. $20-$30.

#3708 Quack Quack Duck with scarf, moving mouth and waddling action. Windup, plush with tin. 4.75 in (12 cm). 1969-1971. Scarcity: 4. $20-$40.

#4162 Quacking Duck with walking, moving head and quacking sound. Windup, plush with plastic. 6 in (15 cm). 1973-1977. Scarcity: 3. $20-$40.

Bugs & Butterflies

(left) **#2177 Lady Bug** with baby on back. Windup, tin and plastic. 4 in (10 cm). 1979-1980s. Scarcity: 3. $20-$40. (right) **#2176 Somersault Lady Bug** with baby on back turns somersaults. Windup, tin and plastic. 4 in (10 cm). 1978-1980. Scarcity: 3. $25-$45.

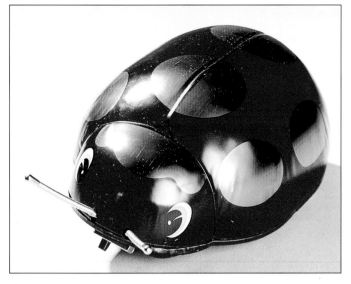

#4032 Friendly Bug with running and tumbling action. Battery operated, tin. 6.75 in (17 cm). 1972-1973. Scarcity: 4. $25-$50.

#4032 Friendly Bug sample. Mock up sample for #4032. 1972.

#1513 Butterfly Ballet with Music. Butterfly on wire moves among flowers on wire from flower pot base containing music box. Windup, tin. 13 in (33 cm). 1955-1960. Scarcity: 5. $75-$150.

#3365 Butterfly Ballet. Butterflies on wires move around flower basket base with music. Windup, tin and plastic. 13 in (33 cm). 1965-1971. Scarcity: 4. $75-$125.

Bugs – Not Pictured

#1742 **B-Z Beetle** with long proboscis. Battery operated, mystery action, tin. 9.5 in (24 cm). 1957-1958. Scarcity: 3. $75-$125.

#1782 **Busy Bizzy Friendly Bug** with antennae. Battery operated, mystery action, tin. 8.75 in (22 cm). 1957-1958. Scarcity: 4. $75-$150.

#1868 **Busy Nosy Bug** with baby ladybug on its back. Battery operated, non-fall, tin. 8 in (20 cm). 1958. Scarcity: 4. $50-$100.

Cats

#3124 **Kitty & Butterfly**. Cat chases butterfly on spring wire. Windup, cellulose acetate and tin. 4.25 in (11 cm). 1962. Scarcity: 6. $30-$60.

#1836 **Happy Time Pussy**. Kitty chasing tin ball on wire rod. Windup, tin. 8 in (20 cm). 1958-1960. Scarcity: 3. $40-$75.

#3100 **My Kitty** with bouncing ball attached to paw by string. Windup, tin and vinyl. 3 in (8 cm). 1961-1973. Scarcity: 2. $25-$50.

#3269 **Butterfly Catcher**. Cat chasing butterfly attached by spring. Windup, tin. 4.5 in (11 cm). 1964-1967. Scarcity: 4. $40-$75.

#4171 **My Kitty** with bouncing ball attached to paw by string. Windup, tin and vinyl. 3 in (8 cm). 1961-1973. Scarcity: 2. $25-$50.

#3698 **Cat & Mouse**. Cat chases "Mighty Mouse" like figure, stops and raises up. Battery operated, plush with plastic. 13 in (33 cm). 1969-1973. Scarcity: 5. $100-$175.

#4235 **R/C Lovely Cat** with wagging tail, turning head, opening mouth and voice. R/C battery operated, plush and plastic. 9 in (23 cm). 1974-1978. Scarcity: 3. $30-$50.

#3781 **Playful Pussy**. Kitty chases ball. Battery operated, mystery action, plush. 12 in (30 cm). 1969-1979. Scarcity: 2. $35-$55.

#3859 **Playsome Cat With Ball**. Cat moves her head while playing with ball on a string. Windup, plastic. 5.5 in (14 cm). 1971-1972. Scarcity: 5. $20-$40.

#4443 **Holiday Cat** with ball in rocking chair. Battery operated, plush and plastic. 9 in (23 cm). 1976-1980s. Scarcity: 2. $25-$45.

#4016 **Pussy Cat - Fancy Cat** with paws moving a ball and mewing sound. Battery operated, mystery action, plush and plastic. 12 in (30 cm). 1972-1973. Scarcity: 4. $35-$75.

#4632 **Playful Kitty** with ball between paws. Battery operated, mystery action, plush and plastic. 15 in (38 cm). 1978-1980s. Scarcity: 2. $30-$60.

Cats 61

#4634 **Mew Mew Pussy**. Kitty walks, moves her head and stops and mews. Battery operated, plush. 9 in (23 cm). 1978-1980s. Scarcity: 2. $20-$40.

#4713 **Kitty Ball** with ball and squeaking voice. Battery operated, mystery action, plush. 8 in (20 cm). 1979-1980s. Scarcity: 2. $20-$35.

#4865 **Walking Kitty**. Kitty walking and mewing. Battery operated, plush. 8 in (20 cm). 1981. Scarcity: 2. $20-$40.

#4704 **Sniffing Kitty** stays low to the ground while moving face and tail. Battery operated, non-fall, plush. 10 in (25 cm). 1979-1980s. Scarcity: 3. $25-$50.

Cats – Not Pictured

#1869 **Mouse Hunter**. Cat chasing mouse attached by rod. Friction, tin. 6 in (15 cm). 1958. Scarcity: 3. $50-$75.

#1880 **Mew-Mew Walking Cat** remote controlled walking kitty. R/C battery operated, plush over tin. 9 in (23 cm). 1958-1962. Scarcity: 6. $60-$100.

#3005 **Walking Kitty**. Windup, plush over tin. 6.5 in (17 cm). 1960. Scarcity: 2. $20-$40.

#3041 **Playful Pussy** chases ball on rod. Battery operated, plush over tin. 10 in (25 cm). 1960-1964. Scarcity: 2. $30-$60.

#3082 **Mew Mew Pussy** walking kitty. Battery operated, plush over tin. 9 in (23 cm). 1960-1964. Scarcity: 6. $60-$100.

#3125 **Walking Kitty**. Battery operated, plush over tin. 9 in (23 cm). 1962-1962. Scarcity: 2. $30-$60.

#3131 **Romping Pussy**. Kitty playing with tin ball between its paws while laying on back. Battery operated, plush and tin. 9.5 in (24 cm). 1961-1962. Scarcity: 4. $50-$100.

#3306 **Playful Pussy** with ball and squeaking sound. Battery operated, mystery action, plush. 12 in (30 cm). 1964-1969. Scarcity: 3. $30-$50.

#3377 **Pussy Cat** with ball on wire. Battery operated, non-fall, tin and plastic. 9.5 in (24 cm). 1965-1967. Scarcity: 2. $25-$45.

#3387 **Walking Pussy** walks with mew sound. R/C battery operated, plush. 9 in (23 cm). 1966-1967. Scarcity: 4. $25-$45.

#3529 **Ball & Kitty**. Cat playing with ball on attached spring. Windup, plastic with tin. 7 in (18 cm). 1967-1969. Scarcity: 2. $20-$30.

#3596 **Pretty Pussy** walking with umbrella. Windup, plastic with tin. 7.5 in (19 cm). 1968-1970. Scarcity: 2. $20-$30.

#3646 **Fumbling Pussy**. Cat rises to feet while fumbling ball. Battery operated, plush. 11 in (28 cm). 1968-1970. Scarcity: 4. $50-$80.

#3786 **Butterfly Catcher**. Kitty chases butterfly suspended on spring wire. Windup, plastic. 7 in (18 cm). 1970-1972. Scarcity: 2. $15-$25.

#3846 **Ball & Kitty**. Kitty plays with ball on spring wire. Windup, plastic. 7 in (18 cm). 1970-1972. Scarcity: 2. $15-$25.

#4060 **Pussy Cat** with moving head, wagging tail and mewing sound. Battery operated, mystery action, plush. 9.75 in (25 cm). 1972-1973. Scarcity: 3. $25-$50.

#4207 **Happy Cat Family**. Balloon waves and cats shake their heads with sound. Battery operated, mystery action, plush and plastic. 12 in (30 cm). 1974-1975. Scarcity: 3. $40-$75.

#4295 **Ball & Kitty** with moving head and arm playing ball. Windup, vinyl. 5.5 in (14 cm). 1975-1976. Scarcity: 3. $20-$35.

Dogs

#2205 **Cheerful Puppy**. Patting the dog while running causes a change in direction. Windup, tin and plastic. 4 in (10 cm). 1980. Scarcity: 3. $20-$30.

Mickie Puppie with spinning ball and moving mouth. Windup, celluloid. 8 in (20 cm). 1930s. Scarcity: 7. *Courtesy of Rex & Kathy Barrett.* $300-$500.

#3034 **Playful Puppy with Caterpillar**. Puppy chases moving caterpillar on base. Battery operated, plush and tin. 7.5 in (19 cm). 1959-1962. Scarcity: 4. $100-$200.

#3437/3782 **Frisky Puppy** shaking a shoe with squeaking sound. Battery operated, mystery action, plush. 9 in (23 cm). 1969-1977. Scarcity: 2. $35-$55.

#3684 **Naughty Dog & Buzzing Bee (Sniffy Dog with Bee)**. Dog chasing bee attached by wire. Nose lights up when bee stings nose. Battery operated, plush with plastic. 13 in (33 cm). 1966-1973. Scarcity: 5. $75-$125.

Puppy & Bee. Puppy with bee on tail. Windup, Celluloid. 3.5 in (9 cm). 1940s. Scarcity: 5. $75-$125.

Dogs   63

#3900 **Shopping Dog**. Two different dogs produced with shopping basket in their mouths. Battery operated, mystery action, plush and plastic. 8.5 in (22 cm). 1971-1972. Scarcity: 5. $30-$50.

#4234 **R/C Lovely Spaniel** with wagging tail, turning head, opening mouth, and voice. R/C battery operated, plush and plastic. 9 in (23 cm). 1974-1980s. Scarcity: 3. $30-$50.

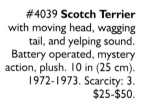

#4039 **Scotch Terrier** with moving head, wagging tail, and yelping sound. Battery operated, mystery action, plush. 10 in (25 cm). 1972-1973. Scarcity: 3. $25-$50.

#4155 **Walking and Yelping Dachshund**. One button causes dog to walk and wag tail, the other to yelp and move his head and mouth. R/C battery operated, plastic. 11.5 in (29 cm). 1973. Scarcity: 7. $75-$150.

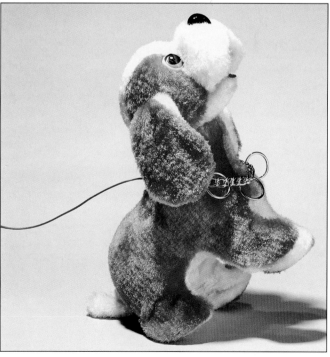

#4303 **Happy Puppy** with walking and stand up action. R/C battery operated, plush. 10 in (25 cm). 1975-1979. Scarcity: 2. $25-$45.

64 *Animals*

#4444 **Holiday Puppy** with baby bottle in rocking chair. Battery operated, plush and plastic. 9 in (23 cm). 1976-1980s. Scarcity: 2. $25-$45.

#4411 **Lucky Spaniel (Red)**. Walking spaniel with wagging tail and shaking head stops and barks. Battery operated, plush. 9 in (23 cm). 1976. Scarcity: 4. $20-$40.

#4656 **Walking Spaniel** walks, stops, and barks by remote control. R/C battery operated, plush. 7 in (18 cm). 1978-1980. Scarcity: 2. $25-$50.

#4470 **Pull Toy Poodle** with chain leash. Pull toy, plush. 9 in (23 cm). 1976-1978. Scarcity: 4. $20-$30.

#4412 **Lucky Spaniel (Brown)**. Walking spaniel with wagging tail and shaking head stops and barks. Battery operated, plush. 9 in (23 cm). 1976. Scarcity: 4. $20-$40.

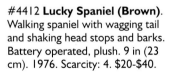

#4490 **Cutie Spaniel**. Walking dog with wagging tail and shaking ears, stops and barks. Battery operated, plush. 12 in (30 cm). 1976-1980s. Scarcity: 2. $25-$50.

#4491 **Lovely Walking Dachshund** with walking, wagging tail, and barking. R/C battery operated, plush. 10 in (25 cm). 1976-1980s. Scarcity: 3. $30-$50.

Dogs 65

#4737 **New Family Puppy**. Dog, with puppies in cart, moves mouth and arm and puppies cry. Battery operated, mystery action, plush and plastic. 12 in (30 cm). 1979-1980. Scarcity: 4. $50-$100.

#4796 **Wee Wee Pekinese Dog** walks, stops, and raises hind leg while barking. Battery operated, plush. 10 in (25 cm). 1980-1980s. Scarcity: 3. $30-$60.

#4798 **Wee Wee Basset Hound** walks, stops, and raises hind leg while barking. Battery operated, plush. 12 in (30 cm). 1980-1980s. Scarcity: 3. $30-$60.

Dogs – Not Pictured

Strutting Puppy with walking action. Windup, plush over tin. 4.75 in (12 cm). 1950s. Scarcity: 3. $30-$50.

#1059 **Shoes Dog**. Dog shaking tin shoe. Windup, tin and celluloid. 6 in (15 cm). 1940s. Scarcity: 4. $75-$150.

#3034b **Puzzled Puppy**. Puppy chases moving mouse in pipe (variation of Playful Puppy with Caterpillar). Battery operated, plush and tin. 7.5 in (19 cm). 1962. Scarcity: 4. $100-$200.

#3197 **Whine Whine Walking Puppy** walking and barking. Battery operated, plush over tin. 8.5 in (22 cm). 1962-1966. Scarcity: 2. $25-$50.

#3262 **Dachshund**. Walking dog with jacket. R/C battery operated, plush. 1964-1964. Scarcity: 3. $40-$75.

#3436 **Shopping Puppy** with basket in his mouth and sound. R/C battery operated, plush and plastic. 9.25 in (23 cm). 1966-1969. Scarcity: 2. $40-$75.

#3584 **Pretty Poodle** walking with umbrella. Windup, plastic with tin. 7.5 in (19 cm). 1967-1969. Scarcity: 2. $20-$30.

#3610 **Scotch Watch**. Dog with Scotch plaid walks then stops to growl. R/C battery operated, plush. 8 in (20 cm). 1968-1970. Scarcity: 3. $30-$50.

#3901 **Charlie with Leash**. Two different dogs produced carrying shoe or bag in their mouths. Dogs are on leash with suction cup. Battery operated, mystery action, plush and plastic. 9.25 in (23 cm). 1971-1973. Scarcity: 3. $30-$50.

#3902 **Fumbling Puppy**. Two different dogs produced that kick around a ball attached by a spring wire. Battery operated, mystery action, plush and plastic. 12 in (30 cm). 1971-1976. Scarcity: 3. $35-$60.

#4050 **Happy Dog Family**. Dog pulling cart with puppy. Moving heads and balloon along with yelping sounds. Battery operated, mystery action, plush with plastic. 12 in (30 cm). 1972-1978. Scarcity: 2. $50-$100.

#4233 **R/C Lovely Terrier** with wagging tail, turning head, opening mouth and voice. R/C battery operated, plush and plastic. 8 in (20 cm). 1974-1977. Scarcity: 3. $30-$50.

#4272 **Lovely Walking Dog** with wagging tail and stand up action. R/C battery operated, plush and plastic. 11 in (28 cm). 1974-1976. Scarcity: 3. $40-$60.

#4304 **Happy Dog** with walking and stand up action. R/C battery operated, plush. 10 in (25 cm). 1975-1980s. Scarcity: 2. $25-$45.

#4471 **Pull Toy Spaniel** with chain leash. Pull toy, plush. 9 in (23 cm). 1976-1978. Scarcity: 4. $20-$30.

#4473 **Pull Toy Dog** on wheels with crying voice. Pull toy, plush. 13.5 in (34 cm). 1976-1977. Scarcity: 4. $15-$30.

#4489 **Cutie Terrier** walking dog with wagging tail and shaking ears, stops and barks. Battery operated, plush. 12 in (30 cm). 1976-1980s. Scarcity: 2. $25-$50.

#4712 **Sniffing Puppy** stays low to the ground while moving face and tail. Battery operated, non-fall, plush. 10 in (25 cm). 1979-1980s. Scarcity: 3. $25-$50.

#7143 **Jumping Puppy**. Different colored puppies hop forward. Windup, plush. 7.5 in (19 cm). 1971-1976. Scarcity: 2. $20-$40.

#7469 **Jumping Puppy** with walking action. R/C battery operated, plush and plastic. 8.5 in (22 cm). 1974-1975. Scarcity: 3. $30-$50.

66   *Animals*

Donkeys

Walking Donkey with walking action. Windup, plush over tin. 5 in (13 cm). 1950s. Scarcity: 4. $30-$50.

#4440 **Pull Toy Donkey** on wheels with crying voice. Pull toy, plush. 13.5 in (34 cm). 1976. Scarcity: 4. $15-$30.

#4483 **Donkey Wagon**. Windup, plastic. 6 in (15 cm). 1977-1978. Scarcity: 2. $15-$30.

Donkeys – Not Pictured

#7243 **Jumping Donkey** hops forward. Windup, plush. 8 in (20 cm). 1972-1976. Scarcity: 2. $20-$40.

Elephants

The Elephant walks on four legs. Windup, plush over tin. 6 in (15 cm). 1950s. Scarcity: 3. *Courtesy of Barbara Moran.* $40-$75.

#1232 **Walking Elephant** walks on all four legs. Windup, plush over tin. 5 in (13 cm). 1950-1960. Scarcity: 2. $40-$80.

#3467/3813 **Hurray the Elephant** walks with roaring sound. R/C battery operated, plush. 11 in (28 cm). 1966-1971. Scarcity: 3. $40-$80.

#3526 **Pretty Elephant** walking with umbrella. Windup, plastic with tin. 7.5 in (19 cm). 1967-1969. Scarcity: 2. $20-$30.

Elephants – Not Pictured

#3297 **Walking Elephant** with blinking eyes and screech sound. R/C battery operated, plush over tin. 1963. Scarcity: 5. $40-$80.

#7222 **Jumping Elephant** hops forward. Windup, plush. 8 in (20 cm). 1972-1976. Scarcity: 2. $20-$40.

Monkeys

#1747 **Roaring Gorilla Shooting Gallery** with tin cork shooting gun. Hitting gorilla in tummy causes him to raise his arms and roar while eyes light. Battery operated, tin. 9.5 in (24 cm). 1957-1958. Later version used rubber tipped dart gun Scarcity: 5. $175-$350.

#1690 **Walking Gorilla Good Time Charley**. Gorilla walks on feet and arms in realistic fashion. Windup, tin and plush. 6 in (15 cm). 1957-1960. Scarcity: 4. $75-$125.

#1811 **Walking Monkey** with cloth jacket walks with long arms. R/C battery operated, tin and plush. 6 in (15 cm). 1958-1961. Scarcity: 3. $40-$80.

68 Animals

#3105 **Monkey on Scooter**. Plush monkey on flat 3-wheel scooter. Windup, tin and plush. 6 in (15 cm). 1961-1962. Scarcity: 2. $25-$50.

#3422 **Mischievous Monkey** teases dog with bone until dog chases monkey up the tree. Battery operated, tin with plastic monkey and leaves. 13 in (33 cm). 1966-1970. Scarcity: 5. $150-$350.

#3077 **Monkey On Tree**. Plush monkey swings on wire stand in a variety of positions. Windup, tin and plush. 7 in (18 cm). 1960-1962. Scarcity: 5. $75-$125.

#3885 **Dandy Chimp "The Mexican."** Chimpanzee that walks with swinging arms. Windup, plush and plastic. 7 in (18 cm). 1971-1973. Scarcity: 3. $25-$50.

#3275 **Shooting Gorilla** raises arms and lights up when shot with dart gun. Battery operated, tin. 9.5 in (24 cm). 1963. Scarcity: 4. $150-$250.

Monkeys – Not Pictured

Jolly Chimp walks on arms. Windup, plush and tin. 6 in (15 cm). 1950s. Scarcity: 3. $30-$50.

Musical Jocko. Monkey playing cymbals. Windup, fur and tin. 8 in (20 cm). 1940s. Scarcity: 3. $30-$50.

#1056 **Tumbling Monkey** with long rotating arms that cause body to tumble. Windup, celluloid. 4 in (10 cm). 1940s. Scarcity: 3. $75-$125.

#1096 **Jim Dandy** chimp with comb and mirror. Windup, celluloid. 5 in (13 cm). 1940s. Scarcity: 4. $75-$150.

#1662 **Climbing Tom Tom Monkey** with monkey climbing tree. Windup, tin. 13 in (33 cm). 1956-1960. Scarcity: 6. $125-$250.

#1767 **Walking Gorilla** with lighted eyes, swinging arms, and roaring sound. R/C battery operated, tin. 8 in (20 cm). 1957. Scarcity: 5. $150-$300.

#4783 **Drummer Monkey** with shaking body while playing drum. Battery operated, plush and plastic. 15 in (38 cm). 1980-1980s. Scarcity: 3. $40-$75.

Rabbits

#1773 **B-Z Rabbit**. Battery operated, tin. 7 in (18 cm). 1958. Scarcity: 4. $75-$150.

#3022 **Telephone Rabbit**. Rabbit on her rocking chair with old time telephone. Battery operated, plush and tin. 9.5 in (24 cm). 1960. Scarcity: 7. *Courtesy of Smith House Toys.* $150-$300.

#3752 **Bunny & Carrot**. Rabbit plays with carrot suspended on wire, stops and raises up. Battery operated, plush with plastic. 13 in (33 cm). 1969-1972. Scarcity: 3. $50-$100.

#3130 **Bunny The Busy Secretary** with bunny working adding machine and answering phone. Battery operated, plush and tin. 8 in (20 cm). 1961-1962. Scarcity: 8. *Courtesy of Smith House Toys.* $250-$500.

#4185 **Bunny** moves forward on irregular wheels. Windup, plastic with tin. 5.75 in (15 cm). 1973. Scarcity: 4. $20-$40.

#4326 **Happy Family Rabbit**. Rabbit pulling cart with bunnies. Moving heads and balloon along with yelping sounds. Battery operated, mystery action, plush with plastic. 12 in (30 cm). 1975-1980s. Scarcity: 2. $50-$100.

Animals

#4667 **Bunny Guitarist** with bunny playing musical guitar. Battery operated, plush with plastic. 13 in (33 cm). 1978-1979. Scarcity: 4. $30-$60.

#8006 **Walking Drummer Rabbit** walks while playing drum. Windup, plush and plastic. 11 in (28 cm). 1980-1980s. Scarcity: 2. $25-$45.

#3060 **Turtle Family**. Walking tortoise connected with two little turtles following. Windup, tin. 14 in (36 cm). 1959-1966. Scarcity: 4. *Courtesy of Barbara Moran*. $50-$100.

(left) #3778 **Lovely Turtle** with walking action. Windup, plastic. 6 in (15 cm). 1970-1973. Scarcity: 2. $10-$20. (right) #3970 **Happy Turtle Family**. Walking turtle with baby turtle following. Windup, plastic. 9 in (23 cm). 1970-1977. Scarcity: 2. $10-$20.

Rabbits – Not Pictured

#3133 **Telephone Rabbit**. Rabbit on seat with modern telephone. Battery operated, plush and tin. 10 in (25 cm). 1960-1962. Scarcity: 6. $150-$300.

#3141 **Jolly Rabbit with Robin**. Rabbit sits with robin on hand. Battery operated, plush and tin. 10 in (25 cm). 1960-1962. Scarcity: 10. $600-$1,200.

#3402 **Lovely Bunny** walks with vibrating action. Windup, plastic with tin. 5.25 in (13 cm). 1965-1971. Scarcity: 2. $20-$30.

#4441 **Holiday Rabbit** with comb and mirror in rocking chair. Battery operated, plush and plastic. 9 in (23 cm). 1976-1979. Scarcity: 3. $30-$50.

#7199 **Jumping Rabbit** hops forward. Windup, plush. 8 in (20 cm). 1972-1978. Scarcity: 2. $20-$40.

Turtles – Not Pictured

#3209 **Papa & Sonny**. Father turtle with son turtle on back. Windup, tin. 5.5 in (14 cm). 1962-1966. Scarcity: 3. $40-$75.

#3602 **Lovely Tortoise** with walking action. Windup, plastic with tin. 6.75 in (17 cm). 1968-1969. Scarcity: 3. $25-$35.

#3641 **Family Tortoise** with baby tortoise and walking action. Windup, plastic with tin. 9.75 in (25 cm). 1968-1969. Scarcity: 2. $25-$35.

Other and Grouped Animals

Turtles

#3054 **Walking Tortoise** walks in realistic manner. Windup, tin. 5.5 in (14 cm). 1960-1966. Scarcity: 4. $40-$75.

Walking Giraffe with leaf in mouth and walking action. Windup, plush over tin. 6 in (15 cm). 1950s. Scarcity: 4. $40-$60.

Other and Grouped Animals 71

#1857 **Lion Target Game** with gun and darts. When target is hit, lion roars and eyes light. Battery operated, tin. 7 in (18 cm). 1958-1960. Scarcity: 6. *Courtesy of Cybertoyz.* $150-$275.

#3090 **Chirping Grasshopper** with moving legs. Battery operated, mystery action, tin. 8.5 in (22 cm). 1960-1962. Scarcity: 6. $100-$200.

#1901 **Bubble Lion**. Lion with cub blows bubbles from mouth. Battery operated, tin. 9 in (23 cm). 1958. Scarcity: 4. $150-$300.

#3790 **Lovely Alligator** with walking action. Windup, plastic. 9 in (23 cm). 1970-1973. Scarcity: 2. $10-$20.

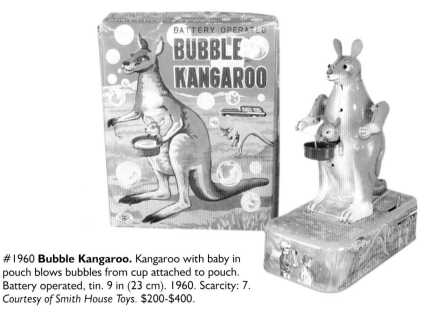

#1960 **Bubble Kangaroo.** Kangaroo with baby in pouch blows bubbles from cup attached to pouch. Battery operated, tin. 9 in (23 cm). 1960. Scarcity: 7. *Courtesy of Smith House Toys.* $200-$400.

Animals

#4269 **Happy Band Trio** with dog, bear, and rabbit playing music and moving. Battery operated, tin, plush and plastic. 11 in (28 cm). 1974-1980s. Scarcity: 4. $150-$300.

(left) #7928 **Playful Panda Bear Family**. Sweet faced animal plays with ball by moving it up and down. Windup, plush and plastic. 7 in (18 cm). 1978-1979. Scarcity: 2. $20-$40. (center) #7914 **Playsome Monkey**. Sweet faced animal plays with ball by moving it up and down. Windup, plush and plastic. 7 in (18 cm). 1978-1979. Scarcity: 2. $20-$40. (right) #7915 **Pretty Rabbit**. Sweet faced animal plays with ball by moving it up and down. Windup, plush and plastic. 7 in (18 cm). 1978-1979. Scarcity: 2. $20-$40.

(left) #7925 **Tumbling Monkey**. Plush animal tumbles. Windup, plush. 6 in (15 cm). 1979. Scarcity: 2. $15-$30. (right) #7926 **Tumbling Bear**. Plush animal tumbles. Windup, plush. 6 in (15 cm). 1979. Scarcity: 2. $15-$30.

(left to right) #7764 **Plump Elephant**. Jumping elephant playing with ball. Windup, plush and plastic. 8.5 in (22 cm). 1976-1979. Scarcity: 2. $20-$40. #7714 **Chick & Ball**. Jumping chick playing with ball. Windup, plush and plastic. 8 in (20 cm). 1976-1980s. Scarcity: 2. $20-$40. #7701 **Plump Rabbit**. Jumping rabbit playing with ball. Windup, plush and plastic. 8.5 in (22 cm). 1976-1980s. Scarcity: 2. $20-$40. #7698 **Plump Doggie**. Jumping dog playing with ball. Windup, plush and plastic. 8.5 in (22 cm). 1976-1980s. Scarcity: 2. $20-$40.

Other Animals – Not Pictured

Pig's Joker. Walking pig with cane. Windup, tin. 3 in (8 cm). 1940s. Scarcity: 4. $75-$125.

#1670 **Walking Animal**. Dog, bear, or elephant walk with moving legs as they are pulled along by leash. Pull toy, plush over tin. 9 in (23 cm). 1956-1960. Scarcity: 4. $25-$50.

#1827 **Musical Mushroom**. Mushroom shaped umbrella spins on base with swinging animals and bell.. Battery operated, tin with celluloid. 12 in (30 cm). 1958-1960. Scarcity: 9. $200-$325.

#1930 **Hungry Sheep** walks, bleats, and eats paper. R/C battery operated, plush and tin. 9 in (23 cm). 1959. Scarcity: 7. $200-$350.

Boats

Toy boats featured in this section mirror the popular boats and ships of the 1950s and 1960s. Open motorboats are listed first, followed by cabin cruisers, ocean going ships, pirate schooners, paddlewheel boats, tugboats, and submarines.

Motorboats

(left to right) **Shot Boat** with crank inertia motor and tin pilot. Friction, tin. 9 in (23 cm). 1958. Scarcity: 5. $100-$200.
#1852 **Champion Boat** with crank inertia motor, visible engine, and twin tin pilots. Friction, tin. 9 in (23 cm). 1958-1960. Scarcity: 5. $125-$250. #3215 **C-15 Boat**. Speedboat 3215 with crank inertia motor, visible engine, and vinyl head pilot. Friction, tin with vinyl. 9 in (23 cm). 1962. Scarcity: 3. $50-$100.

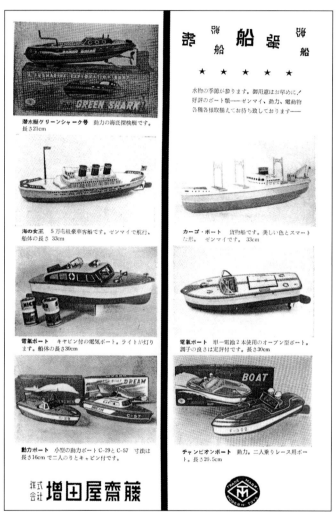

Trade journal ad, April 1958.

#1791 **O-106 Electric Boat with Blinking Light**, spinning prop and pilot. Battery operated, tin. 13 in (33 cm). 1958-1962. Scarcity: 5. $100-$175.

74   Boats

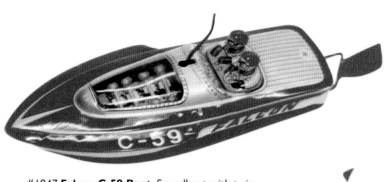

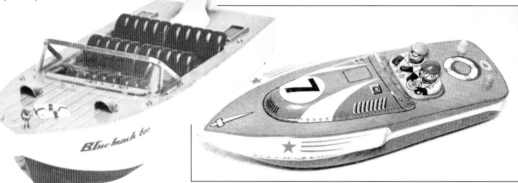

#3589 **Blue Mach Open Boat** with pilot, lights, and spinning prop. Battery operated, plastic and tin. 12.25 in (31 cm). 1968-1972. Scarcity: 4. $100-$175.

#1947 **Falcon C-59 Boat**. Speedboat with twin figures, hand crank, visible moving pistons and spinning prop. Friction, tin. 11 in (28 cm). 1959-1962. Scarcity: 4. $100-$175.

#2653 **Blue Mach 600 Propeller Boat** with large above water propeller for airboat design. Battery operated, tin and plastic. 12.5 in (32 cm). 1968-1973. Scarcity: 4. $125-$200.

#3162 **Seven Sea Boat**. No.7 speedboat with twin pilots and no rudder. Friction, tin. 9.25 in (23 cm). 1962. Scarcity: 6. $75-$150.

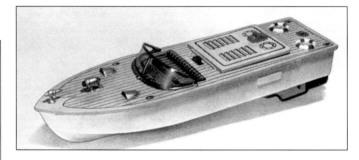

#3566 **Open Boat w/ light** and rotating prop for water use. Battery operated, tin with plastic. 9.5 in (24 cm). 1967-1971. Scarcity: 4. $40-$60.

#3161 **Swift Arrow Boat**. Swift Arrow 8 speedboat with twin pilots, crank motor and rudder. Friction, tin. 10.5 in (27 cm). 1962. Scarcity: 6. *Courtesy of Rex & Kathy Barrett.* $150-$275.

#3266 **S-106 Electric Boat with Blinking Light** with tin driver. Battery operated, tin. 13 in (33 cm). 1962. Similar to #1791. Scarcity: 6. $150-$275.

Cruisers

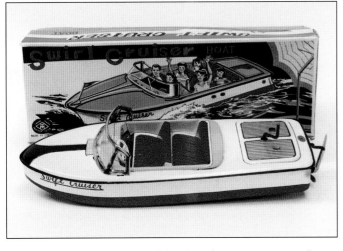

#3166 **Swift Cruiser Boat** with hand crank, spinning prop, and rudder. Friction, tin. 10.5 in (27 cm). 1962. Scarcity: 4. $75-$125.

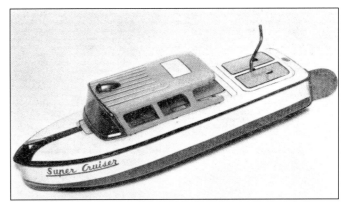

#3165 **Super Cruiser Boat** with hand crank, spinning prop and rudder. Friction, tin. 11 in (28 cm). 1962. Scarcity: 4. $75-$125.

Motorboats – Not Pictured

#1699 **Speed Boat C-55** with hand crank, propeller and rudder. Friction, tin. 11 in (28 cm). 1957-1960. Scarcity: 4. $75-$125.

#1744 **Motor Boat Dream C-57**. Dream boat with lithographed figures on windows. Friction, tin. 7 in (18 cm). 1957-1958. Scarcity: 4. $75-$150.

#1746 **Motor Boat C-29 with 2 Men** with hand crank and turning propeller. Friction, tin. 7 in (18 cm). 1957-1958. Scarcity: 5. $100-$200.

#3563 **Open Boat** with hand crank and rotating prop for water use. Friction, tin with plastic. 9.5 in (24 cm). 1967-1970. Scarcity: 4. $30-$50.

Cabin Cruisers

#3588 **Blue Mach Cabin Boat** with lights and spinning prop. Battery operated, plastic and tin. 12.25 in (31 cm). 1968-1972. Scarcity: 4. $100-$150.

#1714 **Patrol Boat**. O-108 cabin cruiser with propeller and roof light. Battery operated, tin. 13 in (33 cm). 1957-1962. Scarcity: 5. $100-$200.

#1722 **Radicon Boat**. Cabin cruiser with antenna and radio control. R/C battery operated, tin. 18 in (46 cm). 1957-1958. Scarcity: 8. *Courtesy of Smith House Toys.* $250-$450.

Ships

King George V with three stacks, British flags and irregular front wheel for ocean motion. Windup, tin. 19 in (48 cm). 1930s. Scarcity: 9. $1,200-$2,000.

#3176 **Harbor Queen Boat** with spinning prop, engine noise, and flashing lights. Battery operated, tin. 18 in (46 cm). 1962. Scarcity: 6. *Courtesy of Don Bryant.* $200-$400.

Luxury Liner. Boat moves back and forth in irregular pattern between lighthouse and dock. Windup, tin. 12 in (30 cm). 1950s. Scarcity: 6. $250-$375.

#3565 **Torpedo Boat** with hand crank and rotating prop for water use. Friction, tin with plastic. 9.5 in (24 cm). 1967-1970. Scarcity: 4. $30-$50.

#3567 **Cabin Boat w/ light** and rotating prop for water use. Battery operated, tin with plastic. 9.5 in (24 cm). 1967-1971. Scarcity: 4. $40-$60.

Cabin Cruisers – Not Pictured

#1875 **Fire Boat** with hand crank motor and full figured fireman with nozzle on bow. Friction, tin. 13 in (33 cm). 1958. Scarcity: 6. $125-$225.

#2872 **Cabin Boat** with turning prop for real cruising on water. Friction, plastic. 6.5 in (17 cm). 1969-1970. Scarcity: 6. $20-$40.

#3564 **Cabin Boat** with hand crank and rotating prop for water use. Friction, tin with plastic. 9.5 in (24 cm). 1967-1970. Scarcity: 4. $30-$50.

#3568 **Torpedo Boat w/light** and rotating prop for water use. Battery operated, tin with plastic. 9.5 in (24 cm). 1967-1971. Scarcity: 4. $40-$60.

Speed Ship with guns. Windup, tin. 6.75 in (17 cm). 1950s. Scarcity: 5. $100-$200.

Ships 77

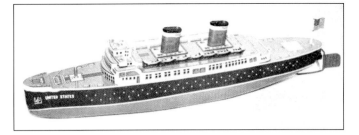

#1642A **Queen of the Sea United States** passenger ship with two stacks, US flag, spinning propeller and rudder. Windup, tin. 14 in (36 cm). 1956-1962. Scarcity: 6. $175-$300.

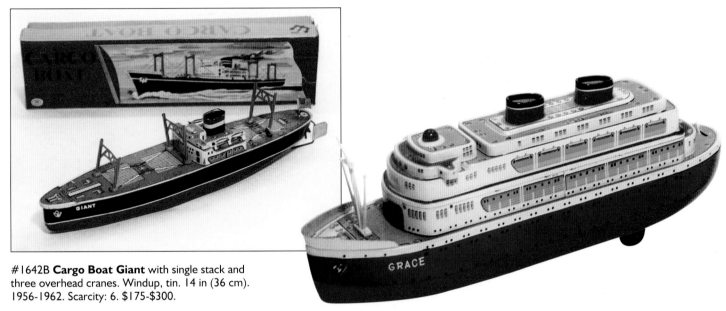

#1642B **Cargo Boat Giant** with single stack and three overhead cranes. Windup, tin. 14 in (36 cm). 1956-1962. Scarcity: 6. $175-$300.

#3382 **Ocean Liner "Grace"** with flashing lights and whistle sound. Battery operated, mystery action, tin. 15.25 in (39 cm). 1965-1969. Scarcity: 7. $275-$525.

#3240 **Queen of the Sea** with flashing lights and deep whistle sound. Battery operated, mystery action, tin. 21.5 in (55 cm). 1962-1967. Scarcity: 6. $200-$400.

78   Boats

Schooners

#3421 **Pirate Ship** with 2-high sail, front sail, rocking action and lighted treasure box. Battery operated, mystery action, tin with plastic. 14 in (36 cm). 1966. Scarcity: 7. $175-$350.

#3503 **Pirate Ship w/Three Sail & Cannon** with 3-high sail, cannon, rocking action, and lighted treasure box. Battery operated, mystery action, tin with plastic. 14 in (36 cm). 1967-1968. Scarcity: 6. $150-$300.

Paddlewheel

#3180 **Whistling Show Boat**. Queen River paddle wheel showboat with forward twin stacks and whistling sound. Battery operated, mystery action, tin. 13.5 in (34 cm). 1961-1964. Scarcity: 3. $125-$250.

#3595 **River Steam Boat** with side paddle wheel, rear twin stacks, smoke, and whistling sound. Battery operated, mystery action, tin with plastic. 13.5 in (34 cm). 1968-1969. Variation of #3200. Scarcity: 5. $125-$250.

#3200 **Show Boat with Whistle & Smoke** has paddle wheel, rear twin stacks, smoke and whistling sound. Battery operated, mystery action, tin with plastic. Hull has rippled stamping at water line to simulate wave effect. 13.5 in (34 cm). 1962-1967. Scarcity: 6. $150-$300.

Tugboats

#3500 **Fire Boat w/Fire Man** with fireman at pump, lights, bell, and pitching motion. Battery operated, mystery action, tin. 14.5 in (37 cm). 1967-1970. Scarcity: 6. $150-$300.

#3290 **Tug Boat Neptune** with whistle, sound, and light. Battery operated, mystery action, tin. 14.5 in (37 cm). 1963-1967. Scarcity: 3. $125-$250.

#3688 **Harbor Tug Boat**. No. 88 with whistling sound, blinking lights, and vinyl head pilot. Battery operated, mystery action, tin with vinyl. 14.5 in (37 cm). 1969-1971. Scarcity: 5. $125-$250.

#3632 **Marine Patrol Boat** with captain standing on deck with large deck mounted siren. Battery operated, mystery action, tin with plastic. 14.5 in (37 cm). 1968. Scarcity: 7. $175-$325.

Submarines

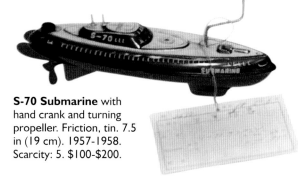

S-70 Submarine with hand crank and turning propeller. Friction, tin. 7.5 in (19 cm). 1957-1958. Scarcity: 5. $100-$200.

#3714 **Ring Smoke Tug Boat**. No.14 with sound and rocking action. Battery operated, mystery action, tin. 14.5 in (37 cm). 1969-1973. Scarcity: 4. $125-$250.

#1743 **Green Shark** submarine with hand crank and turning propeller. Friction, tin. 8 in (20 cm). 1957-1958. Scarcity: 5. $100-$200.

Submarines – Not Pictured

#1481 **Submarine T.M. 300** with hand crank, propeller, forward gun, and flag. Friction, tin. 11 in (28 cm). 1955-1960. Scarcity: 6. $125-$225.

#1517 **Submarine T.M.105** with hand crank, propeller, rear gun and flag. Friction, tin. 14 in (36 cm). 1955-1960. Scarcity: 6. $150-$225.

#1866 **Atomic Submarine A-30** with hand crank friction motor, spinning prop, and missile launcher. Friction, tin. 11 in (28 cm). 1958. Scarcity: 6. $125-$250.

#1872 **Atomic Submarine A-50** with hand crank friction motor and spinning prop. Friction, tin. 8 in (20 cm). 1958. Scarcity: 5. $100-$200.

#3869 **Comical Hop-Up Boat** with blinking light and captain sitting on pop-up smoking funnel. Battery operated, mystery action, plastic. 10.5 in (27 cm). 1971-1973. Scarcity: 4. $35-$75.

Buses

Trade journal ad, May 1964.

#1365 **All State Express** "See America First" bus with roof mounted bullet lights made of rubber. Friction, tin. 8.75 in (22 cm). 1952. Scarcity: 4. $100-$200.

#1657 **Radicon Bus** with grille, single headlights, antenna, and radio control. R/C battery operated, tin. 14 in (36 cm). 1955-1960. Scarcity: 5. $200-$400.

Not only were buses an everyday part of public transportation and an ideal toy subject, they also provided a large surface ideal for a variety of lithographed graphics. Variations in color schemes, passengers in the windows, animals, and business logos could all make the same bus body look totally different. This benefit of reusing the same stamping dies with different lithography also extended into many character toys. Be sure to look at the character toy section for more bus toys illustrated with famous character figures.

Buses

#1666 **Sight Seeing Bus**. All State bus with lithographed passengers in windows. Friction, tin. 14 in (36 cm). 1956-1960. Scarcity: 7. $200-$400.

#1739A **Non-Stop Bus** with lithographed passengers on windows. Battery operated, mystery action, tin. 14 in (36 cm). 1957-1958. Scarcity: 5. $200-$300.

#3081b **Sonicon Bus** with lithographed windows, plastic radar-like antenna and whistle for sound activation. Battery operated, tin. 13.5 in (34 cm). 1966-1967. Scarcity: 5. $150-$300.

#3156A **Sound Bus**. Sightseeing, red and white with lithographed windows, luggage rack, and beep-beep sound. Battery operated, mystery action, tin. 14 in (36 cm). 1961-1963. Scarcity: 7. $150-$300.

#3156B **Sound Bus**. Overland Sightseeing number 800 with whoo-whoo sound. Battery operated, mystery action, tin. 14 in (36 cm). 1964-1969. Scarcity: 5. $150-$300.

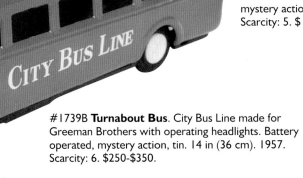

#1739B **Turnabout Bus**. City Bus Line made for Greeman Brothers with operating headlights. Battery operated, mystery action, tin. 14 in (36 cm). 1957. Scarcity: 6. $250-$350.

Buses 83

#3203 **Blue Bird Bus**. Blue Bird Sight Seeing Bus with marker lights and roof vents. Friction, tin. 14 in (36 cm). 1961-1967. Scarcity: 5. $175-$300.

(left) #3420 **Double Decker Bus** with driver and lithographed passengers in windows. Battery operated, mystery action, tin. 13 in (33 cm). 1966-1968. Scarcity: 5. $175-$350. (right) #3392 **Old Fashioned Bus** with driver and lithographed passengers on windows. Battery operated, mystery action, tin. 12.75 in (32 cm). 1966-1968. Scarcity: 4. $150-$300. *Bill & Stevie Weart Collection.*

#3229 **Trolley Bus** Number 3229 with lithographed passengers in windows and double trolley pole. Friction, tin. 14 in (36 cm). 1962-1966. Produced in both an English version and Japanese version. Scarcity: 7. $175-$350.

#3466 **Old Fashioned Steam Coach**. London to Bath coach with horn, side lamps and shaking motion. Battery operated, tin. 14.75 in (37 cm). 1966-1968. Scarcity: 7. $175-$300.

#3483 **Sightseeing Bus**. GM bus with seats and siren sound. Friction, tin. 16 in (41 cm). 1967-1973. Scarcity: 5. $150-$300.

84 Buses

(left) #3703 **Volkswagen Bus "KLM."** VW Transporter KLM. R/C battery operated, tin. 5.5 in (14 cm). 1968-1971. Scarcity: 5. $100-$200.

(right) #3704 **Volkswagen Bus "SAS."** VW Transporter SAS. R/C battery operated, tin. 5.5 in (14 cm). 1968-1971. Scarcity: 5. $100-$200.

#3814 **Sightseeing Bus JNR**. Sightseeing bus of Japanese National Railway system. Friction, tin. 16 in (41 cm). 1970-1973. Scarcity: 5. $150-$300.

#4827 **Panda Bear Bus** with sound, panda bear litho, and Japanese text. Friction, tin. 16 in (41 cm). 1981. Scarcity: 4. $50-$100.

#4173 **Volkswagen Bus Polizei** with flashing light, siren sound and lithographed windows. Battery operated, mystery action, tin and plastic. 10.25 in (26 cm). 1973. Scarcity: 5. $75-$150.

#3536 **Radicon Bus**. Second generation Radicon. GM type bus with marker lights, dual headlights, antenna, and push button radio control. R/C battery operated, tin. 16 in (41 cm). 1967-1970. Scarcity: 4. $175-$350.

Buses – Not Pictured

#1716 **Sight Seeing Bus No. 3** World Line. Friction, tin. 6 in (15 cm). 1957-1960. Scarcity: 5. $60-$125.

#1717 **Queen Bus** double decker London Transport with lithographed passengers. Friction, tin. 6.5 in (17 cm). 1957-1960. Scarcity: 7. $100-$200.

#1867 **Sonicon Bus** with open windows, plastic radar-like antenna, and whistle for sound activation. Battery operated, tin. 13.5 in (34 cm). 1958. Renumbered as #3081 in 1960. Scarcity: 5. $150-$300.

#3081a **Sonicon Bus** with clear dark windows, plastic radar-like antenna, and whistle for sound activation. Blue and white or red and silver. Battery operated, tin. 13.5 in (34 cm). 1960-1967. Scarcity: 5. $150-$300.

#3320 **Animal Bus** with animals lithographed in windows. Friction, tin. 1964. Scarcity: 6. $75-$150.

#3576 **Ambulance Bus**. VW Transporter Ambulance. R/C battery operated, tin. 5.5 in (14 cm). 1967-1972. Scarcity: 5. $100-$200.

#3682 **Volkswagen Bus "Polizei."** VW Transporter with lithographed windows. R/C battery operated, tin. 5.5 in (14 cm). 1968-1972. Scarcity: 5. $100-$175.

#3683 **Volkswagen Bus "Lufthansa."** VW Transporter Lufthansa. R/C battery operated, tin. 5.5 in (14 cm). 1968-1971. Scarcity: 6. $100-$200.

#3699 **Volkswagen Bus "Alitalia."** VW Transporter Alitalia. R/C battery operated, tin. 5.5 in (14 cm). 1968-1972. Scarcity: 6. $100-$200.

#3705 **Volkswagen Bus "Swiss."** VW Transporter Swissair. R/C battery operated, tin. 5.5 in (14 cm). 1968-1971. Scarcity: 6. $100-$200.

#3746 **Volkswagen Bus Police Car** with lithographed personnel, antenna, flashing light and beeping horn. Battery operated, tin with plastic. 10.25 in (26 cm). 1969-1971. Scarcity: 6. $125-$225.

#3747 **Volkswagen Bus Ambulance Car** with lithographed personnel, antenna, flashing light and beeping horn. Battery operated, tin with plastic. 10.25 in (26 cm). 1969-1971. Scarcity: 6. $125-$225.

Cars

Old Timers

In 1953, Ford Motor celebrated its 50th anniversary. This prompted Masudaya to make a miniature replica of a Model-T Ford and several similar types of cars with lever action where the lever resembles the gearshift. When the lever is pulled back and released, the spring tension runs the inertia motor. The product, called "*Oldtimer with lever action,*" was one of the best sellers of the year and shipped mainly to the United States.

#3036 **Old Timers Non Stop Overland** with driver and visible ball moving in engine. "Drive 216,000 miles." Battery operated, mystery action, tin with plastic. 10 in (25 cm). 1959-1962. Scarcity: 4. $125-$175.

#3092 **Old-Timers Assortment No.4,5,6,7,8,9** with lever actuated motor. Lever, tin. 7 in (18 cm). 1960-1968. Scarcity: 3. $40-$75.

#3221 **Sunday Driver (Sundy)** with smoking engine, vinyl head tin driver, and marked "Sundy Driver." Battery operated, mystery action, tin. 10 in (25 cm). 1962-1967. Scarcity: 3. $75-$150.

Passenger Cars

#3716 **Old Fashioned Car, 4 Style Assortment** with lever to wind mechanism. Windup-lever, tin. 5.5 in (14 cm). 1969-1972. Scarcity: 3. $35-$75.

Music Car. Buick with lever actuated music box. Friction, tin. 9.5 in (24 cm). 1953. Scarcity: 4. $100-$200.

#3854 **Hop-Up Old Fashioned Car** with hop up man and radiator cap. Battery operated, mystery action, tin and plastic. 9.25 in (23 cm). 1970-1973. Scarcity: 3. $75-$125.

Lite-It-Up Chevrolet with operating headlights. Battery operated, tin. 9 in (23 cm). 1955. Scarcity: 6. $225-$475.

#3915 **Sunday Rambler** with moving driver, hopping radiator cap, and flashing light. Battery operated, mystery action, tin with plastic. 9.25 in (23 cm). 1971-1972. Scarcity: 3. $50-$100.

Super Buick. 1951 Buick convertible or sedan. Friction, tin. 11 in (28 cm). 1952. Scarcity: 4. $150-$300.

Passenger Cars 87

#1809a **Open Car No.6**. 1957 Chevrolet convertible with tin driver. Friction, tin. 10.75 in (27 cm). 1957-1962. Scarcity: 6. $150-$300.

#3613 **Chevrolet Camaro Sedan Car**. 1967 Chevrolet Camaro. Friction, tin. 11 in (28 cm). 1968-1972. Scarcity: 4. $100-$225.

#4653 **Radicon California Van** picture van with separate radio transmitter to control car moving in circle. R/C battery operated, tin with plastic. 8 in (20 cm). 1978-1979. Scarcity: 4. $100-$200.

#3702 **Chevrolet Camaro Open Car** with vinyl driver. Friction, tin. 11 in (28 cm). 1968-1971. Scarcity: 5. $125-$275.

Ford Car. Windup, tin. 5.5 in (14 cm). 1930s. Scarcity: 8. $300-$600.

Cars

#4666 **Sports Van** with picture roller that shines through roof screen and rear window. Battery operated, mystery action, tin with plastic. 8 in (20 cm). 1978-1980. Scarcity: 5. $200-$400.

#3104 **Daihatsu Midget** with vinyl Midget truck bed cover. Friction, tin with vinyl. 8.75 in (22 cm). 1961-1964. Scarcity: 9. *Hakone Toy Museum Collection.* $800-$1,400.

#3106 **Daihatsu Midget** with Japanese postal service markings and rear hatch. Friction, tin. 8.75 in (22 cm). 1962-1964. Scarcity: 9. *Courtesy of Smith House Toys.* $800-$1,300.

#3336 **Compagno Berlina**. Daihatsu with fender mirrors and antenna. Friction, tin. 10.5 in (27 cm). 1964-1967. Scarcity: 6. $250-$450.

Ford Deluxe. Windup, tin. 8.5 in (22 cm). 1936. Scarcity: 9. $1,000-$1,800.

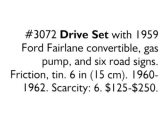

#3072 **Drive Set** with 1959 Ford Fairlane convertible, gas pump, and six road signs. Friction, tin. 6 in (15 cm). 1960-1962. Scarcity: 6. $125-$250.

Passenger Cars   89

62 Ford. 1959 Ford Fairlane 500 hardtop. Friction, tin. 6 in (15 cm). 1962. Scarcity: 5. $85-$185.

Sports Car with Horn. Jaguar XK 120 with tin driver and push button horn. Friction, tin. 7.5 in (19 cm). 1950s. Scarcity: 4. $125-$250.

#3560 **Ford Galaxie 500XL.** 1965 Ford sedan. Friction, tin. 11.5 in (29 cm). 1967-1973. Scarcity: 8. $400-$800.

#4486 **Safari Jeep with Tiger** with engine noise and moving tiger. Battery operated, mystery action, plastic. 9.5 in (24 cm). 1976. Scarcity: 7. $125-$200.

#3561 **Ford Yellow Cab.** 1965 Ford sedan Yellow Cab. Friction, tin. 11.5 in (29 cm). 1967-1973. Scarcity: 8. $450-$950.

90 Cars

Yellow Taxi. 1957 Lincoln taxi with taxi roof and trunk sign. Friction, tin. 8 in (20 cm). 1958. Scarcity: 3. $50-$100.

#3537 **Radicon Benz**. Mercedes Benz 230 SL with antenna and push button radio control. R/C battery operated, tin. 16 in (41 cm). 1967-1973. Scarcity: 5. $200-$400.

#1821 **Lincoln with Siren**. 1957 Lincoln sedan with siren sound. Friction, tin. 12.5 in (32 cm). 1957. Scarcity: 7. $1,200-$2,200.

#3061 **MG Car** with folding windshield. Friction, tin. 8.5 in (22 cm). 1960-1962. Scarcity: 4. $150-$300.

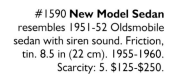

#1590 **New Model Sedan** resembles 1951-52 Oldsmobile sedan with siren sound. Friction, tin. 8.5 in (22 cm). 1955-1960. Scarcity: 5. $125-$250.

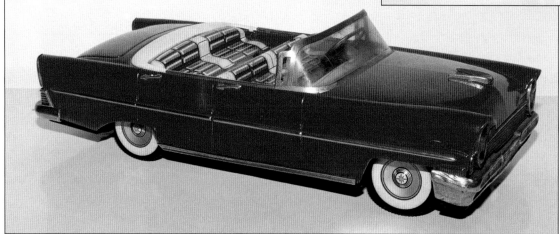

#1821a **Lincoln Open**. 1957 Lincoln convertible. Friction, tin. 12.5 in (32 cm). 1957-1958. Scarcity: 7. $1,200-$2,200.

Passenger Cars 91

#1620 **Convertible with Driver** resembles 1953 Oldsmobile with tin driver. Friction, tin. 6.5 in (17 cm). 1956. Reissued as item #1790. Scarcity: 5. $100-$150.

#1597 **Super Lincoln Sedan Car** that looks like a 1953 Pontiac rather than a Lincoln. Friction, tin. 14 in (36 cm). 1956-1960. Scarcity: 6. $400-$800.

#1774 **Radicon New Sedan**. 1956 Oldsmobile with Radicon radio control and antenna. R/C battery operated, tin. 14 in (36 cm). 1958-1958. Scarcity: 4. $300-$500.

#4699 **Radicon Porsche 930 Turbo -2 channel** with 2 channel radio control. R/C battery operated, plastic. 8.25 in (21 cm). 1979-1980s. Scarcity: 4. $20-$40.

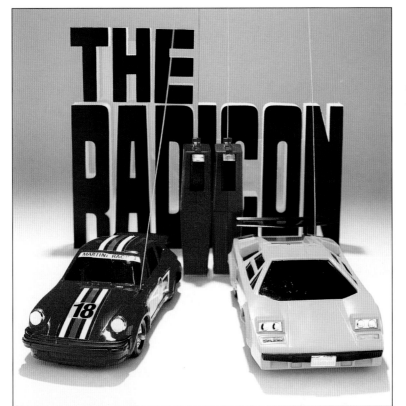

(left) #4638 **Radicon Porsche 930 Turbo** with separate radio transmitter to control car moving in circle. R/C battery operated, plastic. 8.25 in (21 cm). 1978-1980s. Scarcity: 3. $25-$45. (right) #4639 **Radicon Lamborghini Countach** with separate radio transmitter to control car moving in circle. R/C battery operated, plastic. 8.25 in (21 cm). 1978-1980. Scarcity: 3. $25-$45.

92   Cars

Bon Renault. Friction, tin. 7.25 in (18 cm). 1955-1964. Scarcity: 6. $250-$475.

Electric Lucky Car. Renault 4CV sedan with back and forth action. Battery operated, tin. 7.25 in (18 cm). 1950s-1964. Scarcity: 6. $250-$475.

#1607 **Volkswagen** with side lever switch. Battery operated, tin. 7.5 in (19 cm). 1956-1960. Scarcity: 6. $200-$400.

#1707 **Volkswagen** with rear oval window. Friction, tin. 6.5 in (17 cm). 1957-1962. Scarcity: 6. $100-$200.

#3575 **Volkswagen**. R/C battery operated, tin. 5.5 in (14 cm). 1967-1973. Scarcity: 5. $100-$200.

#1624 **New Sunbeam Sports Car**. Friction, tin. 8 in (20 cm). 1956-1960. Scarcity: 5. $150-$300.

Passenger Cars 93

#3583 **Volkswagen (six assorted) Service** with lithographed passengers and six color and graphics variations (Police, Rescue, Fire Chief, PAA, Taxi and rally car). Friction, tin. 6 in (15 cm). 1974-1980s. Scarcity: 3. $25-$50.

#3614 **Volkswagen Sedan** with flashing light and siren sound. Battery operated, mystery action, tin. 10.75 in (27 cm). 1968-1971. Scarcity: 4. $100-$200.

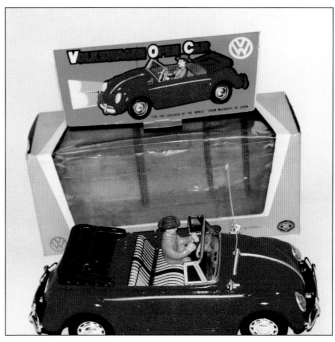

#3757 **Volkswagen Open w/Headlights** with vinyl driver, headlights, and antenna. Battery operated, mystery action, tin. 10.75 in (27 cm). 1969-1972. Scarcity: 7. $250-$450.

#3629 **Volkswagen Taxi** with flashing roof sign and horn. Battery operated, mystery action, tin. 10.75 in (27 cm). 1968-1972. Scarcity: 5. $175-$375.

(left to right) #3934 **Volkswagen Open Car** with vinyl driver, forward/reverse, and steering. R/C battery operated, tin. 10.25 in (26 cm). 1971-1974. Scarcity: 6. $250-$450. #3939 **Volkswagen Taxi** with forward/reverse and steering. R/C battery operated, tin. 10.25 in (26 cm). 1971-1972. Scarcity: 7. $125-$225. #3937 **Volkswagen Sedan** with forward/reverse and steering. R/C battery operated, tin. 10.25 in (26 cm). 1971-1973. Scarcity: 6. $125-$225.

Magic Car with spring bumper, tin driver, and pivoting wheel for changing directions. Friction, tin. 6.75 in (17 cm). 1950s. Scarcity: 4. $40-$75.

Magic Car with spring bumper, tin driver, and pivoting wheel for changing directions. Friction, tin. 5.5 in (14 cm). 1950s. Scarcity: 4. $50-$75.

#4107 **Volkswagen with Detonation** with engine sound. Friction, plastic with tin. 7.5 in (19 cm). 1973. Scarcity: 6. $40-$75.

#1052 **Signal Car** with driver that raises and lowers hand in signaling motion as car turns. Windup, tin. 5 in (13 cm). 1940s. Scarcity: 5. $100-$200.

#4734 **Radicon Volkswagen**. VW 1303 with 4 channel radio control, engine sound, and turn signals. R/C battery operated, plastic. 10 in (25 cm). 1979-1980s. Scarcity: 4. $30-$50.

#1631 **New Open Drive Car No.2** with dog graphics on trunk. Friction, tin. 5.5 in (14 cm). 1956. Scarcity: 3. *Courtesy of Barbara Moran*. $50-$100.

Passenger Cars 95

#1616 **Open Car No. 5** convertible with tin driver. Friction, tin. 9.25 in (23 cm). 1956-1964. Scarcity: 5. $75-$150.

Yellow Taxi No.2 with Yellow Taxi graphics, lithographed passengers, and siren sound. Friction, tin. 5.5 in (14 cm). 1956. Scarcity: 4. $100-$175.

#1895e **Yellow Car**. NYC City Cab variation of Yellow Cab. Friction, tin. 6 in (15 cm). 1958. Scarcity: 6. $100-$175.

#4178 **Amphi-Cat Buggy** with six tires, sound, and turning action. Battery operated, plastic. 9 in (23 cm). 1973. Scarcity: 5. $40-$75.

Passenger Cars – Not Pictured

59 Ford. 1959 Ford Fairlane 500 hardtop. Friction, tin. 8.5 in (22 cm). 1960. Scarcity: 3. $125-$250.

Automoball with Mystery Action. Old fashioned car with ball blowing feature. Battery operated, tin. 10 in (25 cm). 1959. Scarcity: 6. $100-$175.

Buick '59. 1959 Buick hardtop. Friction, tin. 9 in (23 cm). 1960. Scarcity: 3. $100-$200.

Electric Open Car. R/C battery operated, tin. 9 in (23 cm). 1950s. Scarcity: 6. $200-$400.

New Buick. Windup, tin. 4 in (10 cm). 1940s. Scarcity: 4. $75-$150.

New Ford Open. Ford convertible. Friction, tin. 7 in (18 cm). 1955. Scarcity: 5. $65-$125.

#1025 **Baby Pontiac**. 1940s split window model Pontiac. Windup, tin. 3.5 in (9 cm). 1940s. Scarcity: 4. $40-$80.

#1588 **Tiny Car Sedan**. Friction, tin. 5 in (13 cm). 1955-1960. Scarcity: 3. $40-$80.

#1597B **1955 Super Model Open** convertible that looks like a 1953 Pontiac. Friction, tin. 14 in (36 cm). 1956. Scarcity: 6. $400-$800.

#1652 **Baby Sport Car Sedan**. Jaguar sedan. Friction, tin. 5.75 in (15 cm). 1955-1960. Scarcity: 4. $60-$100.

#1667 **Sports Car**. Jaguar XK 120 convertible. Friction, tin. 6.25 in (16 cm). 1956-1958. Scarcity: 5. $100-$175.

#1679 **Station Wagon** based on 1954 Lincoln. Friction, tin. 8.5 in (22 cm). 1956-1960. Scarcity: 4. $75-$150.

#1688 **Zephyr Open Car**. Friction, tin. 6.75 in (17 cm). 1957-1960. Scarcity: 4. $50-$80.

#1705 **Sport Car Open**. Jaguar convertible. Friction, tin. 5.75 in (15 cm). 1957-1960. Scarcity: 5. $100-$175.

#1750 **Lincoln Sedan Car**. 1957 Lincoln with siren sound. Friction, tin. 7.5 in (19 cm). 1957-1958. Scarcity: 4. $100-$200.

#1760 **Open Car No. 50**. Open Jaguar with two tin figures. Friction, tin. 5.75 in (15 cm). 1957-1958. Scarcity: 4. $75-$150.

#1790 **Open Car No. 3**. Oldsmobile convertible with tin driver. Friction, tin. 6.5 in (17 cm). 1957-1958. Same as #1620. Scarcity: 4. $100-$150.

#1794 **Open Car No. 70**. Open Jaguar with two tin figures. Friction, tin. 6 in (15 cm). 1957-1958. Scarcity: 4. $75-$150.

#1809b **Open Car No.6**. 1957 Ford Fairlane convertible with tin driver. Friction, tin. 10.75 in (27 cm). 1958-1964. Scarcity: 6. $150-$300.

#1816 **No.6 Sedan**. 1957 Chevrolet sedan with lithographed figures on windows. Friction, tin. 11 in (28 cm). 1958-1961. Scarcity: 4. $100-$200.

#1891 **Volkswagen Kit**. Oval window Volkswagen as screw assembly kit. Came with body, chassis with two wheels and windup motor assembly with two wheels. Windup, tin. 6 in (15 cm). 1958-1960. Scarcity: 7. $150-$300.

#1895d **Checker Car** with lithographed figures on windows. Friction, tin. 6 in (15 cm). 1958. Scarcity: 4. $100-$175.

#1895f **Desoto Sedan** with lithographed figures on windows. Friction, tin. 6 in (15 cm). 1958. Scarcity: 5. $75-$125.

#1918 **Volkswagen** with popping sound. Friction, tin. 8 in (20 cm). 1958-1962. Scarcity: 3. $100-$200.

#1971 **Sports Car Kit**. Jaguar screw assembly kit. Came with body, four tires, chassis with two wheels and windup motor assembly with two wheels. Windup, tin. 5 in (13 cm). 1960. Scarcity: 7. $75-$125.

#1981 **Buick Station Wagon**. 1959 Buick station wagon. Friction, tin. 9 in (23 cm). 1960. Scarcity: 4. $100-$200.

#2055 **Volkswagen**. Friction, tin. 6.75 in (17 cm). 1977. Scarcity: 5. $50-$75.

#2056 **Mercedes Benz 350 SL**. Friction, tin. 7 in (18 cm). 1977. Scarcity: 5. $40-$60.

#2078 **Aston Martin Car (six assorted)** with lithographed passengers and six color and graphics variations (Police, Ambulance, Fire Chief, Taxi, passenger and rally car). Friction, tin. 7 in (18 cm). 1975. Scarcity: 4. $25-$45.

#2320 **Volkswagen with Spark (six assorted)**. Taxi, Police, Pan Am, Fire Chief, Ambulance and Emergency VW sedan with sparking top lite. Friction, tin with plastic. 5 in (13 cm). 1976-1978. Scarcity: 4. $20-$30.

#3056 **Old and New Car - A Half Century of Progress**. Packaged set with lever wound old-timer car and 9-inch friction 1959 Buick. Friction and lever, tin. 13 in (33 cm). 1960-1962. Scarcity: 5. $150-$275.

#3057 **Daihatsu Midget**. 3-wheel open cab vehicle with vinyl bed canopy. Friction, tin with vinyl. 8.75 in (22 cm). 1960-1962. Scarcity: 9. $800-$1,300.

#3087 **Rolls-Royce** convertible. Friction, tin. 8 in (20 cm). 1960-1962. Scarcity: 6. $250-$450.

#3110 **Halloo Hat** with oversized driver in sports car waving his hat and visible pistons. Battery operated, tin. 9 in (23 cm). 1960-1962. Scarcity: 7. $250-$400.

#3112 **Super Hot Rod** with oversized driver in hot rod sports car waving his arm with V sign and visible pistons. Battery operated, tin. 9 in (23 cm). 1961-1962. Scarcity: 8. $250-$400.

#3189 **Cedric Sedan Car**. 4-door hardtop Nissan. Friction, tin. 7.75 in (20 cm). 1962-1964. Scarcity: 6. $200-$400.

#3359 **R/C Compagno Berlina**. Daihatsu Compagno Berlina sedan with mirrors and antenna. R/C battery operated, tin. 10.5 in (27 cm). 1965-1967. Scarcity: 6. $250-$450.

#3433 **Old Fashioned Car w/Driver** with clicking sound and flashing headlights. Battery operated, non-fall, tin with vinyl. 10.75 in (27 cm). 1966-1967. Scarcity: 6. $125-$225.

#3522 **Sunny 1000**. Nissan Sunny 1000 4-door sedan with fender mirrors. Friction, tin. 11.25 in (29 cm). 1967-1969. Scarcity: 7. $800-$1,200.

#3540 **Sunny 1000**. Nissan Sunny 1000 4-door sedan. R/C battery operated, tin. 11.25 in (29 cm). 1967-1969. Scarcity: 7. $600-$1,000.

#3544/3668 **Volkswagen Sedan**. Red, yellow or blue VW Beetle sedan. Friction, tin. 10.75 in (27 cm). 1967. Replaced by #3668 with detonation. Scarcity: 4. $100-$200.

#3547 **Mercedes Benz 250SL**. Friction, tin. 11 in (28 cm). 1968. Scarcity: 5. $200-$350.

#3554/3669 **Volkswagen Open**. Red, yellow, or blue VW Beetle convertible with motor sound. Friction, tin. 10.75 in (27 cm). 1967. Replaced by #3669 with detonation. Scarcity: 7. $250-$450.

#3574 **Porsche** with forward reverse control. R/C battery operated, tin. 5.5 in (14 cm). 1967-1973. Also produced in Polizei version. Scarcity: 5. $100-$200.

#3581 **Volkswagen Sedan**. Three color variations with lithographed passengers and siren sound. Friction, tin. 6 in (15 cm). 1974-1975. Scarcity: 5. $25-$50.

#3601 **Mercedes Benz 230SL** with steering and flashing lights. R/C battery operated, tin. 15.75 in (40 cm). 1968-1971. Scarcity: 5. $200-$325.

#3657 **Mercedes Benz 230 SL** 2-door coupe. Friction, tin. 13 in (33 cm). 1969. Scarcity: 5. $150-$300.

#3668 **Volkswagen Sedan Car (three colors)**. VW Beetle sedan in red, white, or blue with detonation sound. Friction, tin. 10.75 in (27 cm). 1968-1970. Scarcity: 4. $100-$200.

#3669 **Volkswagen Open Car**. VW Beetle convertible in red, white, or blue with detonation sound. Friction, tin. 10.75 in (27 cm). 1968-1970. Scarcity: 7. $250-$450.

#3756 **Volkswagen Open w/Horn** with vinyl driver, European style horn, and antenna. Battery operated, mystery action, tin. 10.75 in (27 cm). 1969-1970. Scarcity: 8. $250-$450.

#3760 **Lite-Wheel Mercedes**. Mercedes Benz 230 SL with lighted wheels. Battery operated, mystery action, tin. 13 in (33 cm). 1969. Scarcity: 5. $175-$350.

#3944 **Volkswagen Park Avenue** with tin driver and dog passenger. Battery operated, mystery action, tin. 10.25 in (26 cm). 1971. Scarcity: 10. $300-$500.

#4030 **Volkswagen Sedan** with automatic forward and reverse action. Battery operated, tin. 7 in (18 cm). 1972-1973. Scarcity: 5. $75-$150.

Assembly Kits

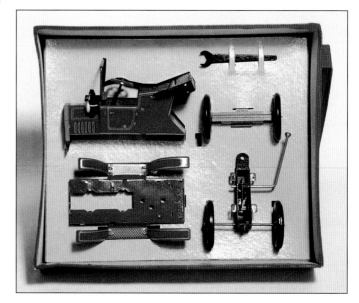

Old Timers Assembly Kit. Assembly kit with body, chassis, wheels, motor, and wrench. Lever, tin. 5.5 in (14 cm). 1960. Scarcity: 7. $50-$125.

Right:
#3334 **Miniature Car Kit** for do-it-yourself assembly consisting of motor, chassis, tires, and body. Both Studebaker and Jaguar included. Windup, tin. 5 in (13 cm). 1964-1966. Scarcity: 8. $200-$300.

#1971B **New Sedan Kit**. Studebaker assembly kit. Came with body, four tires, chassis with two wheels and windup motor assembly with two wheels. Windup, tin. 5 in (13 cm). 1960. Scarcity: 7. $100-$150.

Emergency Cars 97

(left to right) #4158 **Assembly Car Kit Nissan Fairlady**. Assemble yourself kit with rubber tires and windup motor. Windup, plastic with tin. 6 in (15 cm). 1973-1974. Scarcity: 5. $40-$80. #4157 **Assembly Car Kit Toyota Corona Mark II**. Assemble yourself kit with rubber tires and windup motor. Windup, plastic with tin. 6 in (15 cm). 1973-1974. Scarcity: 5. $40-$80. #4156 **Assembly Car Kit Ford Mustang**. Assemble yourself kit with rubber tires and windup motor. Windup, plastic with tin. 6 in (15 cm). 1973-1974. Scarcity: 5. $40-$80.

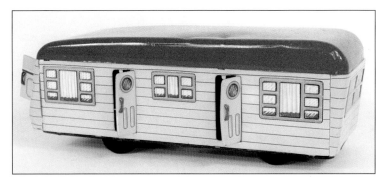

#1651 **Vacationer's Trailer**. 8-inch 1954 Lincoln with 9.5-inch house trailer. Friction, tin. 17.5 in (44 cm). 1956-1960. Scarcity: 5. $125-$250.

Trailers

Vacationer's Trailer No. 3. Lincoln pulling art deco style trailer. Friction, tin. 10 in (25 cm). 1950s. Scarcity: 4. $125-$200.

Emergency Cars

Trade journal ad, February 1968.

98 Cars

Gmen Car. 1940s Pontiac with chrome rear fender skirt and hood mounted machine gun. Windup, tin. 5 in (13 cm). 1950s. Scarcity: 5. *Bill & Stevie Weart Collection*. $100-$200.

Fire Chief Car with lithographed fire figures on windows and press down action. Friction, tin. 5.5 in (14 cm). 1952. Scarcity: 4. $75-$150.

G-Men Car with lithographed figures on windows and press down action. Friction, tin. 7.5 in (19 cm). 1952. Scarcity: 5. $100-$175.

Fire Chief Car No. 5. Press this lithographed fire car down in the rear to activate the inertia friction mechanism. Friction, press action, tin. 9 in (23 cm). 1950s. Scarcity: 4. $125-$225.

#1754 **Fire Dept Car with Light**. Fire Dept. (F.D.) car with lithographed figures on windows and light on roof. Battery operated, tin. 7 in (18 cm). 1957-1958. Scarcity: 3. $50-$100.

Police Squad Car. Press this lithographed police car down in the rear to activate the inertia friction mechanism. Friction, press action, tin. 9 in (23 cm). 1950s. Scarcity: 4. $125-$225.

Emergency Cars 99

#3559 **Ford Fire Chief Car**. 1965 Ford sedan Fire Chief. Friction, tin. 11.5 in (29 cm). 1967-1973. Scarcity: 6. $200-$325.

#3164 **Sonicon Patrol Car**. 1956 Oldsmobile with lithographed windows, roof light, siren, antenna, and whistle sound control. Battery operated, tin. 13.75 in (35 cm). 1962. Photo shows inside mechanism of Sonicon car. Scarcity: 5. $250-$500.

#3486 **Volkswagen Polizei**. Polizei Volkswagen with lithographed passengers on windows. Battery operated, mystery action, tin. 10 in (25 cm). 1967. Scarcity: 6. $200-$400.

#3562 **Ford Police Car**. 1965 Ford Police sedan. Friction, tin. 11.5 in (29 cm). 1967-1973. Scarcity: 6. $200-$325.

#3558 **Ford Ambulance Car**. 1965 Ford sedan ambulance. Friction, tin. 11.5 in (29 cm). 1967-1973. Scarcity: 7. $200-$400.

#3569a **Volkswagen Police Car**. Police Volkswagen with siren and light. Battery operated, mystery action, tin. 10.75 in (27 cm). 1967-1972. Scarcity: 4. $100-$200.

100   Cars

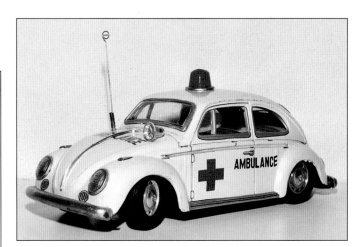

#3645 **Volkswagen "Ambulance"** with flashing top light, antenna, and siren. Battery operated, mystery action, tin. 10.75 in (27 cm). 1968-1972. Scarcity: 5. $100-$200.

#3569b **Volkswagen Polis Car**. Polis Volkswagen variation with siren and light. Battery operated, mystery action, tin. 10.75 in (27 cm). 1967. Scarcity: 4. $100-$200.

#3582 **Volkswagen Polizei** with blue dome light, siren sound and lithographed officers. Friction, tin. 6 in (15 cm). 1974-1977. Scarcity: 4. $25-$50.

#3677 **Ford Police Car**. 1965 Ford Police sedan with roof mounted crank siren. Friction, tin. 11.5 in (29 cm). 1968-1972. Scarcity: 4. $75-$150.

(left) #3740 **Siren Police Jeep** with driver and hood mounted siren activated by rear push button. Battery operated, mystery action, plastic with tin. 9.25 in (23 cm). 1969-1976. Scarcity: 4. $40-$75.
(right) #3741 **Siren Fire Jeep** with driver and hood mounted siren activated by rear push button. Battery operated, mystery action, plastic with tin. 9.25 in (23 cm). 1969-1976. Scarcity: 4. $40-$75.

#3640 **V.W. Riot Squad Car** with large roof mounted siren. Battery operated, mystery action, tin. 10.75 in (27 cm). 1968-1971. Scarcity: 4. $100-$200.

Emergency Cars   101

#3916 **Sonicon Patrol Car (Mercedes)**. Mercedes 230 SL police car with whistle control. Battery operated, tin. 15.75 in (40 cm). 1971-1972. Scarcity: 5. $200-$325.

(left to right) #3936 **Volkswagen Ambulance Car** with flashing light, siren, forward/reverse, and steering. R/C battery operated, tin. 10.25 in (26 cm). 1971-1972. Scarcity: 6. $100-$200. #3935 **Volkswagen Police Car** with flashing light, siren, forward/reverse, and steering. R/C battery operated, tin. 10.25 in (26 cm). 1971-1972. Scarcity: 6. $100-$200. #3938 **Volkswagen Chief Car** with flashing light, siren, forward/reverse, and steering. R/C battery operated, tin. 10.25 in (26 cm). 1971-1973. Scarcity: 6. $100-$200.

#4021 **Highway Patrol Car with Crank Siren**. 1965 Ford with large roof mounted crank siren. Friction, tin with plastic. 11.25 in (29 cm). 1972-1979. Scarcity: 4. $75-$150.

#4025 **Siren Patrol Car**. Japanese market Mercedes Benz 230SL with flashing dome roof light and warning siren sound. Battery operated, mystery action, tin with plastic. 13 in (33 cm). 1972. Scarcity: 5. $150-$300.

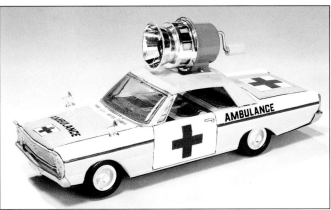

#4036 **Ambulance Car with Crank Siren**. 1965 Ford with large roof mounted crank siren. Friction, tin with plastic. 11.25 in (29 cm). 1972-1976. Scarcity: 4. $75-$150.

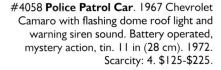

#4058 **Police Patrol Car**. 1967 Chevrolet Camaro with flashing dome roof light and warning siren sound. Battery operated, mystery action, tin. 11 in (28 cm). 1972. Scarcity: 4. $125-$225.

102 Cars

#4058J **Police Patrol Car**. 1967 Chevrolet Camaro with flashing dome roof light and warning siren sound. Battery operated, mystery action, tin. 11 in (28 cm). 1972-1975. Japanese market version shown. Scarcity: 6. $250-$450.

#4068 **Police Patrol Car with New Warning Sound**. Mercedes Benz 230SL with flashing dome roof light and warning siren sound. Battery operated, mystery action, tin with plastic. 13 in (33 cm). 1972-1974. Scarcity: 4. $150-$300.

#4072 **Fire Chief Car with Crank Siren**. 1965 Ford with large roof mounted crank siren. Friction, tin with plastic. 11.25 in (29 cm). 1972-1975. Scarcity: 6. $175-$300.

#4200 **Police Patrol Car** with flashing light and sound. Battery operated, mystery action, plastic with tin. 11.75 in (30 cm). 1973-1975. Scarcity: 4. $75-$150.

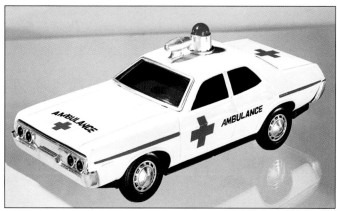

#4254 **Ambulance Car** with flashing light and sound. Battery operated, mystery action, plastic with tin. 11.75 in (30 cm). 1974-1975. Scarcity: 3. $75-$150.

#4294 **Television Patrol Jeep** with driver, siren, and traffic TV screen. Battery operated, mystery action, tin with plastic. 9.25 in (23 cm). 1974-1980s. Scarcity: 4. $50-$100.

Emergency Cars   103

#4424 **Police Patrol Car** with flashing light and sound. Battery operated, mystery action, plastic with tin. 12 in (30 cm). 1976-1977. Scarcity: 3. $75-$150.

#4502 **Police Patrol Car**. Nissan with flashing light and siren sound. Battery operated, mystery action, plastic with tin. 8 in (20 cm). 1976-1978. Scarcity: 4. $75-$125.

#4452 **Highway Patrol Car**. 1967 Chevrolet Camaro with flashing roof lights and warning siren sound. Battery operated, mystery action, tin with plastic. 11 in (28 cm). 1976-1980s. Also sold as New Police Car. Scarcity: 3. $75-$150.

#4579 **Police Patrol Car**. Nissan with flashing multi-color roof lights and warning siren sound. Battery operated, mystery action, tin with plastic. 8 in (20 cm). 1977-1980s. Scarcity: 3. $75-$125.

#4476 **Emergency Police Patrol Car**. 1967 Chevrolet Camaro with spinout action. Battery operated, tin with plastic. 11 in (28 cm). 1976-1980s. Scarcity: 3. $75-$150.

#4598 **Highway Patrol Car** with flashing red and blue lights. Battery operated, mystery action, plastic with tin. 12 in (30 cm). 1977. Scarcity: 3. $75-$150.

#4668 **New Police Car**. 1977 Pontiac Firebird with flashing roof lights and warning siren sound. Battery operated, mystery action, tin with plastic. 10 in (25 cm). 1978-1980s. Scarcity: 3. $50-$100.

#4668 **New Police Car** mock-up sample.

Emergency Cars – Not Pictured

Fire Chief Car with lithographed figures on windows and press down action. Friction, tin. 7.5 in (19 cm). 1952. Scarcity: 5. $75-$125.

#1753 **Police Dept Car with Light**. Police Dept. (P.D.) car with lithographed figures on windows and light on roof. Battery operated, tin. 7 in (18 cm). 1957-1958. Scarcity: 3. $50-$100.

#1882 **No.6 High Way Patrol (1958 Ford)**. 1958 Ford Highway Patrol with lithographed figures on windows. Friction, tin. 11 in (28 cm). 1958-1964. Scarcity: 4. $100-$200.

#1895a **Ambulance Car** with lithographed figures on windows. Friction, tin. 6 in (15 cm). 1958. Scarcity: 4. $60-$100.

#1895b **Police Car** with lithographed figures on windows. Friction, tin. 6 in (15 cm). 1958. Scarcity: 4. $60-$100.

#1895c **Fire Chief Car** with lithographed figures on windows. Friction, tin. 6 in (15 cm). 1958. Scarcity: 4. $60-$100.

#1978 **Buick Highway Patrol**. 1959 Buick police with fender siren and antenna. Friction, tin. 9 in (23 cm). 1960. Scarcity: 4. $100-$175.

#1978A **High Way Patrol Car (1959 Ford)**. 1959 Ford Police hardtop. Friction, tin. 8.5 in (22 cm). 1960-1962. Scarcity: 5. $100-$200.

#1979A **Fire Chief Car (1959 Ford)**. 1959 Ford Fire Chief hardtop. Friction, tin. 8.5 in (22 cm). 1960-1962. Scarcity: 5. $100-$200.

#1980 **Buick Ambulance**. 1959 Buick Ambulance with fender siren and antenna. Friction, tin. 9 in (23 cm). 1960. Scarcity: 4. $100-$175.

#1980A **Ambulance Car (1959 Ford)**. 1959 Ford Ambulance hardtop. Friction, tin. 8.5 in (22 cm). 1960-1962. Scarcity: 5. $100-$200.

#3063 **Police Dept. Car**. 1959 Ford Fairlane 500 Police Dept. Car. Friction, tin. 9 in (23 cm). 1960-1962. Scarcity: 5. $100-$200.

#3107 **Super Patrol Man** with oversized policeman riding in sports car with siren, sound, and flashing light. Battery operated, non-stop, tin. 9 in (23 cm). 1960-1962. Scarcity: 7. $250-$400.

#3111 **Super Fire Man** with oversized fireman riding in sports car with siren, sound, and flashing light. Battery operated, tin. 9 in (23 cm). 1961-1962. Scarcity: 7. $250-$400.

#3167 **Non Fall Police Car** with oversized policeman with vinyl head and red light. Battery operated, non-fall, tin. 9 in (23 cm). 1961-1964. Scarcity: 6. $100-$200.

#3539 **Radicon New Patrol Car**. Van with antenna and push button radio control. R/C battery operated, tin. 13 in (33 cm). 1967-1971. Scarcity: 6. $175-$300.

#3555 **Siren Police Car Cadillac**. Police-Highway Patrol 1966 Cadillac with siren, mirrors, and antenna. Friction, tin. 11.5 in (29 cm). 1967. Scarcity: 6. $150-$250.

#3556 **Benz Polizei**. Polizei Mercedes Benz with European siren sound and flashing light. Battery operated, mystery action, tin. 11 in (28 cm). 1967. Scarcity: 6. $150-$250.

#3621 **Volkswagen "Polizei"** with flashing top light, antenna, and siren. Battery operated, mystery action, tin. 10.75 in (27 cm). 1968. Scarcity: 4. $100-$200.

#3637 **Volkswagen "Chief"** with flashing top light, antenna, and siren. Battery operated, mystery action, tin. 10.75 in (27 cm). 1968-1972. Scarcity: 4. $100-$200.

#3642 **Cadillac Police Car** with siren sound. Friction, tin. 11.5 in (29 cm). 1968. Scarcity: 9. $200-$400.

#3643 **Cadillac Fire Chief Car** with siren sound. Friction, tin. 11.5 in (29 cm). 1968. Scarcity: 9. $200-$400.

#3644 **Cadillac Ambulance Car** with siren sound. Friction, tin. 11.5 in (29 cm). 1968. Scarcity: 9. $200-$400.

#3650 **Cadillac Ambulance Car** with antenna and large roof mounted siren. Battery operated, mystery action, tin. 11.5 in (29 cm). 1968. Scarcity: 9. $200-$400.

#3706 **Ford Fire Chief Car**. 1965 Ford Fire Chief sedan with roof mounted crank siren. Friction, tin. 11.5 in (29 cm). 1969-1972. Scarcity: 6. $175-$285.

#3707 **Ford Ambulance Car**. 1965 Ford Ambulance sedan with roof mounted crank siren. Friction, tin. 11.5 in (29 cm). 1969-1971. Scarcity: 6. $100-$225.

#3733 **Siren Police Car (Mercedes)**. Mercedes 230 SL with vinyl head driver, hood light, and trunk siren activated by push button on driver's hand. Battery operated, mystery action, tin with vinyl. 13 in (33 cm). 1969. Scarcity: 5. $125-$250.

#3734 **Siren Fire Chief Car (Mercedes)**. Mercedes 230 SL with vinyl head driver, hood light and trunk siren activated by push button on driver's hand. Battery operated, mystery action, tin with vinyl. 13 in (33 cm). 1969. Scarcity: 5. $125-$250.

#3762a **Ford Police Car Open**. 1965 Ford convertible with trunk mounted crank siren. Friction, tin. 11.5 in (29 cm). 1969-1971. Scarcity: 5. $125-$225.

#3762b **Ford Police Car Open**. 1965 Ford convertible police variation. Friction, tin. 11.5 in (29 cm). 1969. Scarcity: 6. $125-$250.

#3946 **R/C Benz Patrol Car with Top Light** with steering, flashing light, and forward/reverse action. R/C battery operated, tin. 12.75 in (32 cm). 1971-1973. Scarcity: 4. $150-$300.

#4109 **Fire Chief Car**. 1967 Chevrolet Camaro with flashing dome roof light and warning siren sound. Battery operated, mystery action, tin. 11 in (28 cm). 1973-1975. Scarcity: 4. $100-$225.

#4255 **Fire Chief Car** with flashing light and sound. Battery operated, mystery action, plastic with tin. 11.75 in (30 cm). 1974-1975. Scarcity: 3. $75-$150.

Racing Cars

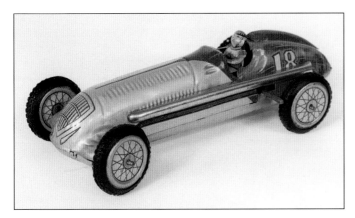

Fairylite Racer with tin driver. Windup, tin. 8.5 in (22 cm). 1950s. Scarcity: 6. $250-$475.

Race Car No. 32 with tin driver. Friction, tin. 5.5 in (14 cm). 1950s. Scarcity: 6. *Courtesy Smith House Toys.* $150-$300.

Hurricane Racer with tin driver. Friction, tin. 13 in (33 cm). 1950s. Scarcity: 8. $300-$600.

Super Sonic Race Car No. 36 with tin driver. Friction, tin. 9.5 in (24 cm). 1950s. Scarcity: 4. $125-$250.

Race Car No. 12 with tin driver. Friction, tin. 4 in (10 cm). 1950s. Scarcity: 5. $75-$150.

#1861 Race Car No.777. Land speed racer with tin driver. Marked Wynn's Friction Proofing. Friction, tin. 9 in (23 cm). 1958. Scarcity: 7. $400-$750.

106   Cars

#3007 **Turn Over Race Car**. 3-wheel racer that flips over and travels in opposite direction. Friction, tin. 8.5 in (22 cm). 1960-1962. Scarcity: 4. $125-$250.

#3301 **Champion Race Car**. Number 301 with driver, sound, and lighted pistons. Battery operated, mystery action, tin and plastic. 18 in (46 cm). 1963-1967. Scarcity: 6. $400-$600.

#3339 **Record Holder Racer** with vinyl head driver, flashing lights, and engine sound. Battery operated, mystery action, tin. 12.5 in (32 cm). 1965-1966. Scarcity: 5. $300-$600.

#3339s **Sample Racer**. Hand painted design mock-up showing possible wheel, graphics and color combinations. See production model #3339 Record Holder. Tin. 13 in (33 cm). 1965. Scarcity: 10.

#3375 **Panther Racer** with driver, sound, and sparking. Friction, tin. 18 in (46 cm). 1965-1966. Scarcity: 7. $400-$600.

Racing Cars 107

#3638 **Champion Racer 38** with lighted pistons, sound and driver. Battery operated, mystery action, tin. 18 in (46 cm). 1968-1972. Scarcity: 6. $400-$600.

#3948 **Jaguar Stunt Car**. Jaguar XKE with roll bars and rollover action. Battery operated, tin with plastic. 9 in (23 cm). 1971-1976. Scarcity: 3. $100-$175.

#3979 **Stunt Car Cadillac #17** with rollover action. Battery operated, tin with plastic. 11.25 in (29 cm). 1972-1975. Photo shows version sold through Sears (#27). Scarcity: 3. $75-$165.

#4288 **Racing Machine with Detonation**. Ford Lotus with engine noise and directional front wheels. Battery operated, tin and plastic. 11.5 in (29 cm). 1974-1976. Scarcity: 4. $125-$350.

#4046ABC **Jaguar, Mercedes Benz & Ford Mustang Stunt Car** with rollover action. Battery operated, tin with plastic. 7 in (18 cm). 1972-1973. Scarcity: 4. $75-$125.

#4359a **Racing Team Car**. 1965 Ford Galaxie with Wild Boar racing graphics. Friction, tin. 11.5 in (29 cm). 1975. Scarcity: 10. No price found.

#4359b **Racing Team Car**. 1965 Ford Galaxie with Thunder Flame racing graphics. Friction, tin. 11.5 in (29 cm). 1975. Scarcity: 10. No price found.

#4359c **Racing Team Car**. 1965 Ford Galaxie with Star Dust racing graphics. Friction, tin. 11.5 in (29 cm). 1975. Scarcity: 10. No price found.

Racing Cars 109

#4728 **Big Key Racer** with big key for on off and lever for direction. Battery operated, plastic. 8.5 in (22 cm). 1979-1980s. Scarcity: 2. $15-$30.

#4682 **Stunt Cars** with roll bars and turnover action. Battery operated, plastic with tin. 8 in (20 cm). 1978-1979. Scarcity: 4. $30-$60.

Racing Cars – Not Pictured

#1694 **Speed Race Car**. Friction, tin. 6 in (15 cm). 1957-1960. Scarcity: 5. $75-$150.

#1720 **Speed Race Car No**. 20 with two tin figures. Friction, tin. 7 in (18 cm). 1957-1958. Scarcity: 6. $125-$225.

#1758 **Race Car No. 25**. Open sports car with tin driver. Friction, tin. 5.75 in (15 cm). 1957-1958. Scarcity: 4. $60-$125.

#1759 **Race Car No. 59**. Open sports car with two tin figures. Friction, tin. 5.75 in (15 cm). 1957-1958. Scarcity: 4. $75-$150.

#1871 **Speed Race Car No.270** with twin race figures. Battery operated, tin. 6.5 in (17 cm). 1958. Scarcity: 4. $75-$125.

#1886 **Champion Racer Set - 3 Car Assorted**. Sold as set of three numbered (#7, 8, 9) cars with twin tin figures. Friction, tin. 8 in (20 cm). 1958-1960. Scarcity: 4. $100-$175.

#2063 **World Racers (Set of 3)** with vinyl head drivers and Lotus, Porsche, or Ferrari racing motifs. Friction, tin. 9.25 in (23 cm). 1975-1978. Also sold as free wheeling version #2117. Scarcity: 5. $50-$100.

#2064 **Racing Car (Set of 3)** with vinyl head drivers and three different racing motifs. Free wheeling, tin. 7 in (18 cm). 1975-1976. Also sold as free wheeling version #2335. Scarcity: 5. $40-$75.

#2065 **Race Car (Set of 3)** with three different racing motifs. Free wheeling, tin. Sold in poly bag. 6.75 in (17 cm). 1975-1976. Scarcity: 3. $20-$40.

#2669 **Race Car 6, 8, 11** in three assorted colors. Windup, tin and plastic. 5 in (13 cm). 1979-1980. Scarcity: 4. $15-$25.

#2819 **Hot-Rod Race car**. Windup, plastic. 7 in (18 cm). 1969-1971. Scarcity: 4. $20-$30.

#3287 **Race Car Seven Star**. Number 7 rocket-like racer with seven stars, driver, and sound. Friction, tin. 12.75 in (32 cm). 1964-1966. Scarcity: 5. $250-$450.

#4031 **Jaguar** XKE race car with steering and forward/reverse. R/C battery operated, tin. 10 in (25 cm). 1972. Scarcity: 5. $65-$135.

#4593a **Rally Car No.8** with engine noise. Battery operated, mystery action, plastic with tin. 8 in (20 cm). 1977-1979. Scarcity: 4. $30-$60.

#4593b **Rally Car No.12** with engine noise. Battery operated, mystery action, plastic with tin. 8 in (20 cm). 1977-1979. Scarcity: 4. $30-$60.

Platform Base Toys

Not Pictured

#1682 **Magic Crossroads (Platform Base)**. Windup car and bus travel dual figure-8 highway without hitting each other. Windup, tin. 18 in (46 cm). 1956-1960. Scarcity: 6. $75-$150.

#1702 **Round High Way (Platform Base)** with vehicles traveling on highway and through tunnel by vibrating action. Battery operated, tin. 24 in (61 cm). 1957-1960. Scarcity: 5. $75-$150.

Character Toys

Toys based on movie, television, or story characters have always sparked the imagination of the little kids who played with them—or the even bigger kids who collect them. This section focuses on the many character-based toys produced by Masudaya. Disney toys are listed first, followed by other character toys including Japanese market characters.

Licensing

There have been many toys produced in Japan based on popular copyrighted characters. To use the likeness of one of these characters required permission or a license by the owner of the copyright. Early licensing was usually obtained by the importers such as George Borgfeldt. For example, if Masudaya produced a Disney toy, the US importer obtained the licensing or authorization to produce the toy. When Marx came to Japan in the 1950s, they had the licensing rights for many character toys.

In 1959, Disney started Walt Disney Japan for the purpose of licensing direct. Of course, Disney eventually opened a theme park in Japan in 1983 where Japanese market only Disney toys were sold. Masudaya became the largest licensee of Disney toys in Japan and maintained that role for many years.

Today, licensing is a cornerstone of the global toy business. While it has become a key profit contributor for the companies who own the licenses, it also has become very expensive for the toy manufacturers.

Disney

Trade journal ads, September 1962, September 1964, and April 1968.

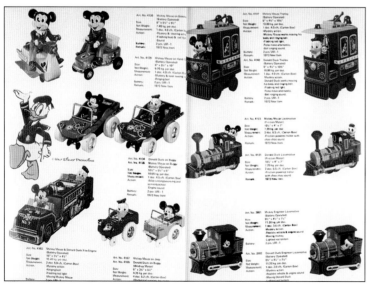

Disney items shown in 1973 catalog.

Disney 111

Running Mickey on Pluto with Mickey on Pluto attached to tin wheel. Windup, celluloid and tin. 6 in (15 cm). 1940s. Scarcity: 9. *Courtesy of Bertoia Auctions.* $1,600-$3,500.

Donald Race Car with Donald Duck head driver. Windup, tin and celluloid. 3 in (8 cm). 1940s. Scarcity: 7. *Courtesy of Smith House Toys.* $175-$300.

Mickey Race Car with Mickey Mouse head driver. Windup, tin and celluloid. 3 in (8 cm). 1940s. Scarcity: 7. *Courtesy of Smith House Toys.* $175-$300.

(top) #2340 **Disney Living Set D-2** with Mickey and Donald refrigerator, Mickey stove and Donald sink. Includes utensils. Tin. 14 in (36 cm). 1966-1970. Scarcity: 8. $200-$400.
(bottom) #2341 **Disney Miniature Living Set D-3** with Snow White refrigerator and kitchen range, sink and washing machine illustrating Seven Dwarfs. Includes utensils. Tin. 14 in (36 cm). 1966-1970. Scarcity: 8. $275-$400.

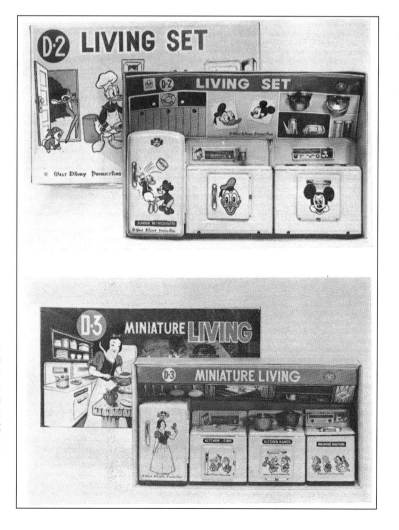

112   Character Toys

#2342 **Disney Washing Machine**. Donald Duck washing machine with hose. Windup, tin. 7 in (18 cm). 1966-1970. Scarcity: 7. $100-$175.

#2343 **Disney Refrigerator with Foods**. Junior Refrigerator with Bambi and plastic fruits. Tin. 6.5 in (17 cm). 1966-1973. Scarcity: 6. $75-$125.

#2454 **Disney's Miniature Living Set D-9** with Bambi refrigerator, Mickey kitchen range, and Donald kitchen sink. Includes utensils. Tin. 10.25 in (26 cm). 1966-1972. Scarcity: 8. $225-$375.

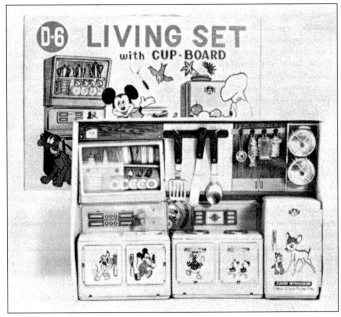

#2367 **Disney Living Set with Cup-Board D-6** with Bambi refrigerator, Donald and Nephews stove, Mickey and Pluto sink, cupboard, and utensils. Box serves as kitchen walls. Tin. 16 in (41 cm). 1966-1969. Scarcity: 7. $250-$375.

#2372 **Disney Refrigerator** with Mickey Mouse opening handle. Tin. 8 in (20 cm). 1966-1970. Scarcity: 7. $100-$175.

#2447 **Disney's Washing Machine**. Automatic washer with Donald Duck and rubber hose. Windup, tin. 4.5 in (11 cm). 1966. Scarcity: 9. $100-$175.

#2492 **Disney Kitchen D-2000** consisting of Donald windup washer, Nephew stove, and Mickey refrigerator. Tin. 19 in (48 cm). 1966-1969. Scarcity: 8. $225-$450.

#2480 **Disney's Tea Set D-80**. 14 piece Lady and the Tramp tea pot, cups and saucers. Tin. 1966-1973. Scarcity: 7. $100-$200.

Art. No. 2899 Disney Barber Shop

(left) #2899 **Disney Barber Shop** with barbering accessories. Paper and plastic. 12 in (30 cm). 1970-1973. Scarcity: 6. $100-$200. (below) #2514 **Disney Doctor Set**. Japanese market Mickey Mouse doctor set with miniature medical supplies. Paper with plastic. 15 in (38 cm). 1966-1977. Scarcity: 7. $30-$60.

Art. No. 2514 Disney Doctor Set

#2493 **Disney's Family Kitchen Set D-800**. 21 piece Disney illustrated kitchen set. Tin. 1966-1972. Scarcity: 8. $100-$150.

114 Character Toys

#2721 **Disney Living Corner 3 pieces D-650** with Disney character stove, sink, and short refrigerator. Tin. 16.75 in (43 cm). 1968-1970. Scarcity: 7. $225-$350.

#3316 **Disney Electric Line** with pantograph and lithographed Disney figures in windows. Friction, tin. 10.5 in (27 cm). 1964. Scarcity: 8. *Courtesy of Smith House Toys.* $400-$650.

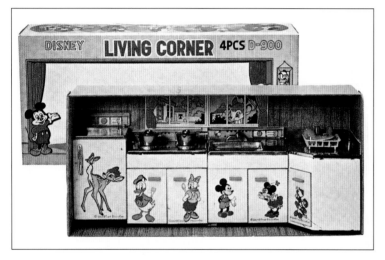

#2722 **Disney's Living Corner 4 pieces D-900** with Disney character corner cabinet, stove, sink, and short refrigerator. Tin. 20 in (51 cm). 1968-1970. Scarcity: 7. $275-$400.

Art. No. 3399 Wan Wan Bus   *Art. No. 4286 Disney Bus*

(left)
#3399 **Wan Wan Bus - Lady and the Tramp** with Lady and the Tramp lithographed figures. Friction, tin. 14 in (36 cm). 1965-1975. Scarcity: 4. $125-$250. (right) #4286 **Disney Bus** with lithographed Disney characters and Japanese text. Friction, tin. 16 in (41 cm). 1974-1979. Scarcity: 6. $200-$400.

Art. No. 2951

Art. No. 7816   *Art. No. 7758*

(left to right) #2951 **Disney Alarm Clock** with three different Disney scenes available (Mickey, Bambi or Dalmatians). Windup, plastic. 3.25 in (8 cm). 1970-1980s. Scarcity: 3. $20-$40. #7816 **Disney Alarm Clock** with Dalmatian, Bambi, or Donald Duck character illustrations. Windup, plastic. 4.5 in (11 cm). 1977-1980s. Scarcity: 4. $20-$40.
#7758 **Disney Alarm Clock** with Donald Duck face. Windup, plastic. 4 in (10 cm). 1976-1979. Scarcity: 4. $10-$30.

Disney   115

#3531 **Donald Duck** with rocking walking action. Windup, vinyl with tin. 5.75 in (15 cm). 1967-1968. Scarcity: 5. $30-$60.

#3609 **Disney Mickey Mouse Racer**. Racer with vinyl Mickey Mouse head and sound. Friction, tin with vinyl. 12.75 in (32 cm). 1968-1969. Scarcity: 6. $250-$450.

#3636 **Disney Winnie The Pooh** with rocking walking action. Windup, vinyl with tin. 5.25 in (13 cm). 1968-1969. Scarcity: 4. $30-$50.

(left) #3586 **Mickey Mouse** with rocking walking action. Windup, vinyl with tin. 6.25 in (16 cm). 1968-1971. Scarcity: 5. $30-$60. (right) #3612 **Disney Dumbo** with rocking walking action. Windup, vinyl with tin. 5.5 in (14 cm). 1968-1971. Scarcity: 5. $30-$60.

#3607 **Donald Duck Car**. Racer with vinyl Donald head and quack-quack sound. Friction, tin with vinyl. 12.75 in (32 cm). 1968-1969. Scarcity: 6. $250-$450.

#3981 **Mickey Engineer Locomotive** with sound, lighted lantern and Mickey Mouse. Battery operated, mystery action, tin with plastic. 9.5 in (24 cm). 1972-1975. Scarcity: 4. $75-$150.

116 Character Toys

#3982 **Donald Duck Engineer Locomotive** with sound, lighted lantern and Donald Duck. Battery operated, mystery action, tin with plastic. 9.5 in (24 cm). 1972-1975. Scarcity: 4. $75-$150.

#3988 **Disney Bus** with Disney characters and siren sound. Friction, tin. 16 in (41 cm). 1972-1973. Scarcity: 6. $125-$250.

#4123 **Mickey Mouse Locomotive** with choo choo sound. Friction, tin and plastic. 10.5 in (27 cm). 1973. Scarcity: 6. $75-$125.

#4128 **Mickey Mouse On Scooter** with Mickey, flashing lights, sound and steering action. Battery operated, mystery action, tin with plastic. 9.5 in (24 cm). 1973-1976. Scarcity: 5. $150-$300.

#4129 **Mickey Mouse On Hand Car** with Mickey, flashing lights, sound, and steering action. Battery operated, mystery action, tin with plastic. 10 in (25 cm). 1973-1976. Scarcity: 4. $150-$300.

Disney 117

(left) #4135 **Mickey Mouse On Buggy** with Mickey, sound, and bouncing-spinning action. Battery operated, mystery action, plastic with tin. 10.75 in (27 cm). 1973-1979. Scarcity: 4. $100-$200. (right) #4134 **Donald Duck On Buggy** with Donald, sound, and bouncing-spinning action. Battery operated, mystery action, plastic with tin. 10.75 in (27 cm). 1973-1979. Scarcity: 4. $100-$200.

#4131 **Donald Duck Locomotive** with choo choo sound. Friction, tin and plastic. 10.5 in (27 cm). 1973. Scarcity: 6. $75-$125.

#4148 **Donald Duck Trolley** with bell, flashing light, and moving poles. Donald moves while pulling bell. Battery operated, mystery action, plastic and tin. 10.75 in (27 cm). 1973-1975. Scarcity: 5. $100-$200.

#4147 **Mickey Mouse Trolley** with bell, flashing light, and moving poles. Mickey moves while pulling bell. Battery operated, mystery action, plastic and tin. 10.75 in (27 cm). 1973-1975. Scarcity: 5. $100-$200.

#4167 **Mickey Mouse On Jeep** with Mickey and bouncing-spinning action. Windup, plastic with tin. 6 in (15 cm). 1973-1975. Scarcity: 5. $50-$100.

#4168 **Donald Duck On Buggy** with Donald and bouncing-spinning action. Windup, plastic with tin. 6 in (15 cm). 1973-1975. Scarcity: 5. $50-$100.

#4183 **Mickey Mouse & Donald Duck Fire Engine** with flashing light and ringing bell. Battery operated, mystery action, tin with plastic. 16 in (41 cm). 1973-1975. Scarcity: 6. $200-$400.

#4169 **Mickey Mouse Gyrocopter** with Mickey Mouse pilot and whirling blade. Windup, plastic with tin. 10.5 in (27 cm). 1973-1976. Scarcity: 4. $40-$75.

#4170 **Donald Duck Gyrocopter** with Donald Duck pilot and whirling blade. Windup, plastic with tin. 10.5 in (27 cm). 1973-1976. Scarcity: 4. $40-$75.

#4301 **Mickey Mouse Go-Cart** with lit visible pistons moving up and down. Battery operated, mystery action, plastic. 9 in (23 cm). 1975. Scarcity: 6. $100-$200.

Disney 119

#4564b **Disney Scooter Donald** with Donald and ringing bell. Windup, plastic with tin bell. 6 in (15 cm). 1977-1979. Scarcity: 5. $50-$100.

#4564a **Disney Scooter Mickey** with Mickey and ringing bell. Windup, plastic with tin bell. 6 in (15 cm). 1977-1979. Scarcity: 5. $50-$100.

#4571 **Mickey Mouse Musical Car** with musical sound and Mickey Mouse driver. Friction, plastic. 7.5 in (19 cm). 1977-1979. Scarcity: 6. $50-$100.

(left) #4581 **Mickey Mouse Police Car** with Mickey Mouse, sound and flashing light. Battery operated, mystery action, plastic with tin. 10.5 in (27 cm). 1977-1980s. Scarcity: 4. $50-$100. (right) #4582 **Donald Duck Fire Chief Car** with Donald Duck, sound and flashing light. Battery operated, mystery action, plastic with tin. 10.5 in (27 cm). 1977-1980s. Scarcity: 4. $50-$100.

(left) #4673 **Donald Duck Piston Race Car** with visible pistons and automatic forward and spin-out action. Battery operated, plastic with tin. 9 in (23 cm). 1978-1980s. Scarcity: 5. $75-$150. (right) #4674 **Mickey Mouse Piston Race Car** with visible pistons and automatic forward and spin-out action. Battery operated, plastic with tin. 9 in (23 cm). 1978-1980s. Scarcity: 5. $75-$150.

120   Character Toys

(left) **#7231/7579 Bambi Coffee Set (5 place settings)**. 5 place setting of coffee cups ands saucers with pot and spoons. Tin. 1973-1975. Scarcity: 6. $50-$70. (right) **#7230/7578 Bambi Coffee Set (3 place settings)**. 3 place setting of coffee cups and saucers with pot and spoons. Tin. 1973-1975. Scarcity: 6. $30-$50.

#4744 **Disney Bus** with 101Dalmations litho and Japanese text. Friction, tin. 16 in (41 cm). 1979. Scarcity: 6. $175-$350.

#7465 **Disney Animals - 4 assorted** with shaking heads. Windup, vinyl. 5.5 in (14 cm). 1974-1976. Scarcity: 2. $15-$25.

#7082 **Mickey Mouse & Donald Duck Fire Engines** with ringing bell and moving ladder. Windup, plastic. 7 in (18 cm). 1971-1980s. Scarcity: 3. $20-$40.

(left) #7084 **Disney Family Party Set D-750**. 18 piece child's dish set with Disney characters. Tin. 1971-1972. Scarcity: 7. $35-$75. (right) #7085 **Disney Family Party Set D-1000**. 20 piece child's dish set with Disney characters. Tin. 1971-1972. Scarcity: 7. $40-$80.

Disney 121

#7765 **Mickey Mouse & Donald Duck Race Cars** with engine sound. Friction, plastic. 7 in (18 cm). 1976-1980. Scarcity: 3. $20-$40.

#8541 **Mickey Mouse Mercedes Benz.** Pullback, tin with plastic windshield and Mickey. 6.25 in (16 cm). 1986. Scarcity: 2. $50-$95.

#7950 **Disney Drummer Dumbo** shaking body and playing drum. Windup, plush. 9 in (23 cm). 1979-1980s. Scarcity: 4. $25-$50.

#2433 **Disney's Junior Cooking Set D-11**. Cooking pots and pans with Disney characters. Tin. 1966-1967. Scarcity: 7. $75-$150.

#2434 **Disney's Junior Coffee Set D-12**. Miniature coffee pot with cups and saucers illustrated with Lady and the Tramp. Tin. 1966-1967. Scarcity: 7. $75-$125.

#2435 **Disney's Family 32 Piece Kitchen Set D-1000**. 32 piece Disney illustrated dinnerware set. Tin. 1966-1968. Scarcity: 6. $100-$200.

#2494 **Donald Duck Kitchen Set D-600** with combination stove and sink, doll, utensils and accessories. Tin. 1966. Scarcity: 9. $250-$350.

#2495 **Disney' Dinner Set D-50**. 18 piece Cinderella illustrated dinnerware set. Tin and plastic. 1966. Scarcity: 9. $100-$150.

#2757 **New Disney Doctor Set** with play doctor's tools with Disney characters. Paper with plastic. 1968-1970. Scarcity: 4. $30-$50.

#3226 **Disney Electric Train Set**. Twin car set with track pantograph, and lithographed Disney characters. Windup, tin. 1962-1964. Scarcity: 8. $350-$700.

#3255 **Disney's Dalmatian Bus** with lithographed windows featuring 101 Dalmatians. "101 Dalmatians" (in Japanese) on roof, "Walt Disney's Dalmatian Bus" on side. Friction, tin. 16 in (41 cm). 1962-1964. Scarcity: 5. $200-$400.

#3620 **Mickey Mouse The Cyclist** with ringing bell. Windup, plastic with tin bell. 5.5 in (14 cm). 1968-1969. Scarcity: 6. $40-$60.

#3947 **Donald Tricycle**. Donald Duck pedaling on tricycle with revolving spinner. Windup, vinyl, tin and plastic. 7.75 in (20 cm). 1971. Scarcity: 8. $100-$175.

#7054 **Disney Roulette**. Japanese market Disney roulette game. Plastic. 1971-1972. Scarcity: 6. $35-$60.

#7766 **Mickey Mouse & Donald Duck Dump Trucks** with ringing bell and dump action. Windup, plastic. 7 in (18 cm). 1976-1980s. Scarcity: 3. $20-$40.

#8555 **Mickey Mouse Space Ship**. Saucer shaped ship with Mickey at controls, Disney figure lithography and lights outside dome. Battery operated, tin and plastic. 8.25 in (21 cm). 1980s. Scarcity: 4. $75-$125.

Disney – Not Pictured

#2322 **Disney's Cooking Set No.1**. Ten piece cooking set with Disney characters. Tin. 1966-1969. Scarcity: 5. $30-$50.

#2368 **Disney's D7 Cooking Set No.2**. 16 piece cooking set with Disney characters. Tin. 1966. Scarcity: 9. $75-$125.

Japanese Characters

Trade journal ad, 1970.

Boats

Great Mazinger Motor Boat with directional and removable outboard engine for water play. Windup, plastic. 7.5 in (19 cm). 1977. Scarcity: 4. $30-$50.

Mirrorman Water Scooter with Mirrorman figure on submarine shaped water scooter. Windup, plastic. 9 in (23 cm). 1970s. Scarcity: 4. $30-$50.

Buses

Great Mazinger Bus with Mazinger graphics. Friction, tin. 16 in (41 cm). 1970s. Scarcity: 7. $200-$350.

Japanese Characters 123

Kamen Rider Amazon Bus with Kamen Rider graphics. Friction, tin. 14 in (36 cm). 1975. Scarcity: 5. $150-$250.

Robocon Bus with TV show Robocon graphics. Friction, tin. 16 in (41 cm). 1970s. Scarcity: 6. $175-$300.

Kamen Rider V3 Bus with Kamen Rider graphics. Friction, tin. 16 in (41 cm). 1974. Scarcity: 6. $175-$300.

Buses – Not Pictured

Oba-Q Bus with Oba-Q (Obake no Q-Taro) ghost graphics. Friction, tin. 16 in (41 cm). 1969. Scarcity: 7. $200-$400.

Cars & Racers

#3794 Keroyon Bus with Keroyon frog graphics. Friction, tin. 14 in (36 cm). 1968. Scarcity: 7. $200-$350.

Getta Robo Car with Crank Siren. 1965 Ford with large roof mounted crank siren. Friction, tin with plastic. 11 in (28 cm). 1970s. Scarcity: 7. $150-$275.

Great Mazinger Car with Crank Siren. 1965 Ford with large roof mounted crank siren. Friction, tin with plastic. 11 in (28 cm). 1970s. Scarcity: 7. $150-$275.

(left) **Space Boy Soran Bus** with Soran character graphics. Friction, tin. 14 in (36 cm). 1967. Scarcity: 8. $300-$600. (right) **Space Boy Soran Rocket Car**. Red, white, and blue space car with vinyl Soran driver, squirrel, and sound. Friction, tin. 12.75 in (32 cm). 1966. Scarcity: 9. $800-$1,400.

#4023 Kamen Rider On Comic Jeep or Buggy with bouncing-spinning action. Windup, mystery action, plastic with tin. 6 in (15 cm). 1972. Scarcity: 6. $100-$175.

#4422 **Robocon Comic Car** with wacky singing sound. Friction, plastic. 7.5 in (19 cm). 1976. Scarcity: 6. $75-$125.

Kamen Rider X Car with Crank Siren. 1965 Ford with large roof mounted crank siren. Friction, tin with plastic. 11 in (28 cm). 1970s. Scarcity: 7. $150-$275.

#3057A **Tetsujin 28 Daihatsu Midget** with Tetsujin 28 graphics on vinyl truck cover. Friction, tin with vinyl. 8.75 in (22 cm). 1960s. Scarcity: 9. $1,200-$1,800.

Spectreman Series toys (**Spectreman** and **Giant Rah**).

(left) **Kotetsu Jeeg Musical Car** with musical sound. Friction, plastic. 7.5 in (19 cm). 1970s. Scarcity: 6. $75-$125.
(right) **Kotetsu Jeeg Racer** with vinyl head Kotetsu Jeeg, corresponding graphics and stabilizer bar. Friction, tin and plastic. 12.5 in (32 cm). 1970s. Scarcity: 7. $200-$300.

Japanese Characters 125

#4017 **Spectreman Baby Car** with Spectreman driver and ringing bell. Windup, plastic with tin. 5.5 in (14 cm). 1972. Scarcity: 6. $40-$75.

Kamen Rider Race Car with engine sound. Friction, plastic. 7 in (18 cm). 1970s. Scarcity: 6. $50-$75.

Great Mazinger Racer with vinyl head Great Mazinger, corresponding graphics, and stabilizer bar. Friction, tin with plastic. 12.5 in (32 cm). 1970s. Scarcity: 7. $200-$400.

Great Mazinger Race Car with engine sound. Friction, plastic. 7 in (18 cm). 1970s. Scarcity: 6. $50-$75.

Kamen Rider Amazon Racer with vinyl head Amazon Rider, corresponding graphics and stabilizer bar. Friction, tin and plastic. 12.5 in (32 cm). 1970s. Scarcity: 7. $200-$400.

126 *Character Toys*

#4445 **Kamen Rider Stronger Racer** with vinyl head Kamen Rider Stronger, corresponding graphics, and stabilizer. Friction, tin and plastic. 12.5 in (32 cm). 1976. Scarcity: 7. $200-$400.

Kamen Rider V3 Racer with vinyl head Kamen Rider V3, corresponding graphics, and stabilizer bar. Friction, tin and plastic. 12.5 in (32 cm). 1970s. Scarcity: 7. *Courtesy of Smith House Toys.* $200-$400.

Cars & Racers – Not Pictured

Astroboy Rocket Car. Blue, white, and red space car 7 with vinyl Astroboy driver and sound. Friction, tin. 12.75 in (32 cm). 1966. Scarcity: 9. $1,200-$2,000.

Gekko Kamen (Moonlight Mask) Race Car with engine sound. Friction, plastic. 7 in (18 cm). 1970s. Scarcity: 6. $50-$75.

Giant Rah Baby Car with Rah driver and ringing bell. Windup, plastic with tin. 5.5 in (14 cm). 1972. Scarcity: 6. $40-$75 *(shown in group photo of Spectreman Series toys, page 124).*

Jumborg Ace Racer with vinyl head Jumborg Ace, stabilizer bar, and corresponding graphics. Friction, tin and plastic. 12.5 in (32 cm). 1970s. Scarcity: 8. $250-$475.

Kamen Rider X Racer with vinyl head Kamen Rider X, corresponding graphics, and stabilizer bar. Friction, tin and plastic. 12.5 in (32 cm). 1970s. Scarcity: 7. $200-$400.

#3751 **Keroyon Frog Baby Car** with Keroyon frog driver and ringing bell. Windup, plastic with tin. 5.5 in (14 cm). 1969-1970. Scarcity: 6. $40-$75.

Keroyon Race Car with engine sound. Friction, plastic. 7 in (18 cm). 1970s. Scarcity: 6. $50-$75.

Mirrorman Car. 1967 Chevrolet Camaro convertible with vinyl Mirrorman and corresponding graphics. Friction, tin with vinyl. 11 in (28 cm). 1970s. Scarcity: 7. $250-$400.

Mirrorman On Comic Jeep with bouncing-spinning action. Windup, mystery action, plastic with tin. 6 in (15 cm). 1975. Scarcity: 5. $75-$125.

Mirrorman Racer with vinyl head Mirrorman, corresponding graphics, and stabilizer bar. Friction, tin and plastic. 12.5 in (32 cm). 1970s. Scarcity: 7. $200-$400.

Oba-Q Car Green and yellow space car with vinyl Oba-Q driver and sound. Friction, tin. 12.75 in (32 cm). 1969. Scarcity: 7. $250-$400.

Oba-Q Car (Reverse Direction). Red and blue car with vinyl Oba-Q driver facing reverse direction. Friction, tin. 12.75 in (32 cm). 1970s. Scarcity: 7. $250-$400.

Oba-Q Race Car with vinyl head Oba-Q driver and P-Chan sitting on hood of race car. Friction, tin and vinyl. 13 in (33 cm). 1970. Scarcity: 8. $400-$700.

Paaman Car with vinyl Paaman driver and sound. Friction, tin. 12.75 in (32 cm). 1969. Scarcity: 7. $250-$400.

Robocon Race Car with engine sound. Friction, plastic. 7 in (18 cm). 1970s. Scarcity: 6. $50-$75.

Space Boy Soran Race Car # 5 with vinyl head Soran driver. Friction, tin and vinyl. 13 in (33 cm). 1966. Scarcity: 9. $500-$1,000.

Tiger Mask Baby Car with Tiger Mask driver and ringing bell. Windup, plastic with tin. 5.5 in (14 cm). 1970. Scarcity: 6. $40-$75.

Tiger Mask Race Car with engine sound. Friction, plastic. 7 in (18 cm). 1970. Scarcity: 6. $50-$75.

Ultra Jack Racer with vinyl head Ultraman, corresponding graphics and stabilizer bar. Friction, tin and plastic. 12.5 in (32 cm). 1970s. Scarcity: 6. $150-$250 *(shown in Vinyl Figures - Windups section, page 134).*

Ultraman Race Car with engine sound. Friction, plastic. 7 in (18 cm). 1970s. Scarcity: 6. $50-$75.

Ultraman Taro Racer with vinyl head Ultraman Taro, corresponding graphics, and stabilizer bar. Friction, tin with plastic. 12.5 in (32 cm). 1970s. Scarcity: 7. $200-$400.

Japanese Characters

Fire Engines

Ultra Jack Fire Engine with ringing bell. Windup, plastic with tin. 5.5 in (14 cm). 1973. Scarcity: 5. $40-$75.

Fire Engines – Not Pictured

Mirrorman Fire Engine with ringing bell. Windup, plastic with tin. 5.5 in (14 cm). 1973. Scarcity: 5. $40-$75.

Norakuro Fire Engine with ringing bell. Windup, plastic with tin. 5.5 in (14 cm). 1973. Scarcity: 6. $40-$75.

Locomotives

Not Pictured

Space Boy Soran Locomotive with Soran and space graphics. Friction, tin. 16 in (41 cm). 1966. Scarcity: 9. $200-$400.

Motorcycles & Scooters

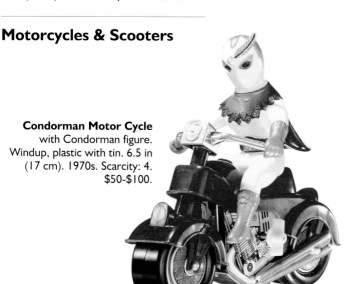

Condorman Motor Cycle with Condorman figure. Windup, plastic with tin. 6.5 in (17 cm). 1970s. Scarcity: 4. $50-$100.

Kamen Rider Amazon Motor Cycle with Amazon Kamen Rider figure. Windup, plastic with tin. 6.5 in (17 cm). 1970s. Scarcity: 6. $75-$150.

Oba-Q on 3-Wheel Scooter. Vinyl Oba-Q ghost on flat 3-wheel scooter. Windup, tin and vinyl. 6 in (15 cm). 1967. Scarcity: 6. $125-$225.

Gekko Kamen (Moonlight Mask) on Scooter with tin Gekko Kamen with vinyl head. Battery operated, tin and plastic. 9 in (23 cm). 1970s. Scarcity: 9. $500-$1,000.

Motorcycles & Scooters – Not Pictured

Great Mazinger Motor Cycle with Great Mazinger figure. Windup, plastic with tin. 6.5 in (17 cm). 1970s. Scarcity: 6. $75-$150.

Kikaida Side-Machine with Kikaida figure and side car. Plastic. 12 in (30 cm). 1973. Hawaii version came packaged with red guitar (+100%). Scarcity: 8. $150-$300.

Kikaida Side-Machine (small size) with Kikaida figure and side car. Plastic. 8 in (20 cm). 1973. Scarcity: 5. $50-$100.

Keroyon on 3-Wheel Scooter. Vinyl Keroyon on flat 3-wheel scooter. Windup, tin and vinyl. 6 in (15 cm). 1967. Scarcity: 6. $100-$200.

Paaman on 3-Wheel Scooter. Vinyl Paaman on flat 3-wheel scooter. Windup, tin and vinyl. 6 in (15 cm). 1968. Scarcity: 6. $125-$225.

P-Chan (Oba-Q) on 3-Wheel Scooter. Vinyl P-Chan (Oba-Q's sister) ghost on flat 3-wheel scooter. Windup, tin and vinyl. 6 in (15 cm). 1968. Scarcity: 6. $125-$225.

#4039 **Kamen Rider Rocket Racer** with vinyl head Kamen Rider. Friction, tin and vinyl. 13 in (33 cm). 1972. Scarcity: 7. $200-$400.

Rocket Racers & Space Vehicles

Gekko Kamen (Moonlight Mask) Rocket Racer with vinyl head Gekko Kamen. Friction, tin with plastic. 13 in (33 cm). 1970s. Scarcity: 7. $200-$400.

Mirrorman Rocket Racer with vinyl head Mirrorman. Friction, tin with plastic. 13 in (33 cm). 1970s. Scarcity: 7. $200-$400.

Oba-Q Rocket Racer with vinyl head Oba-Q. Friction, tin and vinyl. 7 in (18 cm). 1969. Scarcity: 7. $100-$175.

#4373 **Great Mazinger Rocket Racer** with vinyl head Great Mazinger. Friction, tin with plastic. 13 in (33 cm). 1976. Scarcity: 7. $200-$400.

#4753 **Spectreman X-15 Saucer** with vinyl Spectreman and lithographed characters on saucer. Battery operated, mystery action, tin and vinyl. 7 in (18 cm). 1972. Scarcity: 7. $250-$500.

Japanese Characters

Rocket Racers – Not Pictured

Getta Dragon Rocket Racer with vinyl head Getta Dragon G. Friction, tin and vinyl. 13 in (33 cm). 1970s. Scarcity: 7. $200-$400.

Getta Robo Rocket Racer with vinyl head Getta Robo. Friction, tin and vinyl. 13 in (33 cm). 1970s. Scarcity: 7. $200-$400.

Kamen Rider V3 Rocket Racer with vinyl head Kamen Rider V3. Friction, tin and vinyl. 13 in (33 cm). 1974. Scarcity: 7. $200-$400.

Kamen Rider X Rocket Racer with vinyl head Kamen Rider X. Friction, tin and vinyl. 13 in (33 cm). 1975. Scarcity: 7. $200-$400.

Mirrorman Rocket Racer (small) with vinyl head Mirrorman. Friction, tin with plastic. 7 in (18 cm). 1970s. Scarcity: 7. $150-$250.

#3998 Spectreman Rocket Racer with vinyl head Spectreman. Friction, tin and vinyl. 13 in (33 cm). 1972. Scarcity: 7. $200-$400 *(shown in group photo of Spectreman Series toys, page 124)*.

Tiger Mask Rocket Racer with vinyl head Tiger with cape. Friction, tin and vinyl. 13 in (33 cm). 1970. Scarcity: 7. $200-$400.

Tanks

#3926 **Norakuro Tank** with vinyl Norakuro dog on top of rotating turret. Friction, tin with vinyl. 7.5 in (19 cm). 1970. Scarcity: 7. $200-$350.

Space Boy Soran Tank with vinyl Soran and squirrel on top of rotating turret. Friction, tin with vinyl. 7.5 in (19 cm). 1966. Scarcity: 8. $250-$400.

Tricycles

Getta Robo I Tricycle with ringing bell. Windup, plastic with tin. 5 in (13 cm). 1970s. Scarcity: 6. $40-$80.

Kamen Rider Amazon Tricycle with ringing bell. Windup, plastic with tin. 7 in (18 cm). 1970s. Scarcity: 6. $40-$80.

(left) **Kamen Rider II Tricycle** with ringing bell. Windup, plastic with tin. 7 in (18 cm). 1970s. Scarcity: 6. $40-$80. (right) **Keroyon Frog The Cyclist** with ringing bell. Windup, plastic with tin. 5.5 in (14 cm). 1968-1972. Scarcity: 6. $40-$80.

Character Toys

Kamen Rider Tricycle with ringing bell. Windup, plastic with tin. 7 in (18 cm). 1970s. Scarcity: 6. $40-$80.

Silver Kamen Tricycle with ringing bell. Windup, plastic with tin. 7 in (18 cm). 1970s. Scarcity: 6. $40-$80.

#3992 **Spectreman Tricycle** with ringing bell. Windup, plastic with tin. 5.5 in (14 cm). 1971. Scarcity: 6. $40-$80 *(shown in group photo of Spectreman Series toys, page 124)*.

Tiger Mask Tricycle with ringing bell. Windup, plastic with tin. 5.5 in (14 cm). 1970. Scarcity: 6. $40-$80.

Ultra Taro Tricycle with ringing bell and cycle shield. Windup, plastic with tin. 7.5 in (19 cm). 1970s. Scarcity: 6. $40-$80.

Vinyl Figures – Light-ups

Not Pictured

Inazuman with light in head. Battery operated, vinyl. 7.5 in (19 cm). 1974. Scarcity: 7. $35-$100.

Kamen Rider V3 with light in head. Battery operated, vinyl. 7 in (18 cm). 1970s. Scarcity: 7. $35-$100.

Keiji-K Robot Detective with light in head. Battery operated, vinyl. 7 in (18 cm). 1970s. Scarcity: 8. $50-$150.

Mazinger-Z with light in head. Battery operated, vinyl. 7 in (18 cm). 1970s. Scarcity: 8. $50-$150.

Red Baron with light in head. Battery operated, vinyl. 7 in (18 cm). 1970s. Scarcity: 7. $35-$100.

Tiger-7 with light in head. Battery operated, vinyl. 7 in (18 cm). 1970s. Scarcity: 7. $35-$100.

Ultraman Taro with light in head. Battery operated, vinyl. 6.5 in (17 cm). 1970s. Scarcity: 7. $35-$100.

Vinyl Figures – Puppets

#4413 **Robocon Tricycle** with ringing bell. Windup, plastic with tin. 7 in (18 cm). 1976. Scarcity: 6. $40-$80.

Kamen Rider X Tricycle with ringing bell. Windup, plastic with tin. 5 in (13 cm). 1970s. Scarcity: 4. $30-$60.

Great Mazinger Karate puppet with arms that hit quickly when activated internally. Vinyl. 10 in (25 cm). 1976. Scarcity: 6. $40-$75.

Tricycles – Not Pictured

Gekko Kamen Tricycle with ringing bell. Windup, plastic with tin. 7 in (18 cm). 1970s. Scarcity: 6. $40-$80.

Norakuro Tricycle with ringing bell. Windup, plastic with tin. 5.5 in (14 cm). 1968. Scarcity: 6. $40-$80.

Oba-Q Tricycle with ringing bell. Windup, plastic with tin. 5.5 in (14 cm). 1968. Scarcity: 6. $40-$80.

Rainbowman Tricycle with ringing bell and cycle shield. Windup, plastic with tin. 7.5 in (19 cm). 1970s. Scarcity: 6. $40-$80.

Japanese Characters   131

Kamen Rider Amazon Karate puppet with arms that hit quickly when activated internally. Vinyl. 10 in (25 cm). 1976. Scarcity: 6. $40-$75.

Bakulah (Spectreman) articulated posing figure. Vinyl. 9 in (23 cm). 1972. Scarcity: 7. $50-$125.

Vinyl Figures – Posing

Cat. #7180 **Spectreman Series Characters**. 1972. (left to right) Front row: Midoran, Thunder-gai. Middle row: Mogunechudon, Spectreman, Dr. Gori, Giant Rah. Back row: Zeron, Nezubirdon, Dustman, Gokinazaurus, Kurama Nikuras.

Baronsaurus (Spectreman) articulated posing figure. Vinyl. 9 in (23 cm). 1972. Scarcity: 7. $50-$125.

Dustman (Spectreman) articulated posing figure. Vinyl. 9 in (23 cm). 1972. Scarcity: 7. $50-$125.

132 Character Toys

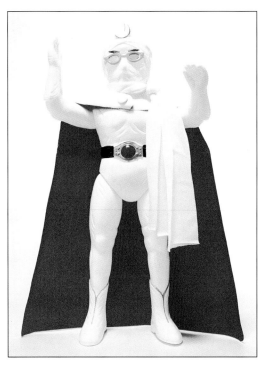

#7228 **Gekko Kamen** articulated posing figure. Vinyl. 10 in (25 cm). 1972. Scarcity: 5. $35-$100.

Mogunechudon (Spectreman) articulated posing figure. Vinyl. 9 in (23 cm). 1972. Scarcity: 7. $50-$125.

Nezubirdon (Spectreman) articulated posing figure. Vinyl. 9 in (23 cm). 1972. Scarcity: 7. $50-$125.

#7201 **Spectreman** articulated posing figure. Vinyl. 17 in (43 cm). 1972. Scarcity: 8. $100-$250.

Kurama Nikuras (Spectreman) articulated posing figure. Vinyl. 9 in (23 cm). 1972. Scarcity: 7. $50-$125.

Thunder-gai (Spectreman) articulated posing figure. Vinyl. 9 in (23 cm). 1972. Scarcity: 7. $50-$125.

Pictured in group photo of Spectreman Series Characters (page 131)

Dr. Gori (Spectreman) articulated posing figure. Vinyl. 9 in (23 cm). 1972. Scarcity: 7. $50-$125.

Giant Rah (Spectreman) articulated posing figure. Vinyl. 9 in (23 cm). 1972. Scarcity: 7. $50-$125.

Gokinazaurus (Spectreman) articulated posing figure. Vinyl. 9 in (23 cm). 1972. Scarcity: 7. $50-$125.

Midoran (Spectreman) articulated posing figure. Vinyl. 9 in (23 cm). 1972. Scarcity: 7. $50-$125.

Spectreman articulated posing figure. Vinyl. 9 in (23 cm). 1972. Scarcity: 5. $35-$100.

Zeron (Spectreman) articulated posing figure. Vinyl. 9 in (23 cm). 1972. Scarcity: 7. $50-$125.

Vinyl Figures – Sparklers

Not Pictured

Denjin Zaboga with sparking in head and push-up sparkler mechanism. Sparkler, vinyl. 7 in (18 cm). 1970s. Scarcity: 7. $35-$100.

Getta Robo 1 with sparking in head and push-up sparkler mechanism. Sparkler, vinyl. 7 in (18 cm). 1970s. Scarcity: 7. $35-$100.

Great Mazinger with sparking in head and push-up sparkler mechanism. Sparkler, vinyl. 7 in (18 cm). 1976. Scarcity: 7. $35-$100.

Kamen Rider X with sparking in head and push-up sparkler mechanism. Sparkler, vinyl. 7 in (18 cm). 1970s. Scarcity: 6. $35-$100.

Ultraman Leo with sparking in head and push-up sparkler mechanism. Sparkler, vinyl. 7 in (18 cm). 1970s. Scarcity: 7. $35-$100.

Vinyl Figures – Swimming

Silver Kamen with Water Motor with removable water motor swimming capsule. Battery operated, vinyl. 9 in (23 cm). 1973. Scarcity: 4. $30-$60.

(left) **#7227 Gokinazaurus (Spectreman)** with pull string talking voice box on back. Talker, vinyl. 13 in (33 cm). 1972. Has been re-issued ($50). Scarcity: 5. $50-$200. (center) **#7212 Spectreman** with pull string talking voice box on back. Talker, vinyl. 14 in (36 cm). 1972. Has been re-issued ($50). Scarcity: 6. $75-$250. (right) **#7226 Kamen Rider** with pull string talking voice box on back. Talker, vinyl. 14 in (36 cm). 1972. Has been re-issued ($50) Scarcity: 6. $100-$300.

Vinyl Figures – Talking

#7753 Condorman with pull string talking voice box on back. Talker, vinyl. 15 in (38 cm). 1976. Scarcity: 6. $75-$200.

#7561 Great Mazinger with pull string talking voice box on back. Talker, vinyl. 15 in (38 cm). 1976. Scarcity: 6. $75-$250.

Jumborg Ace with pull string talking voice box on back. Talker, vinyl. 14.5 in (37 cm). 1972. Scarcity: 10. $1,000-$1,800.

#7751 **Zaboga Stronger** with pull string talking voice box on back. Talker, vinyl. 14.5 in (37 cm). 1976. Scarcity: 8. $125-$400.

Flying Attack Human with pull string talking voice box on back. Talker, vinyl. 14 in (36 cm). 1970s. Scarcity: 9. $175-$500.

Getta Robo 1 with pull string talking voice box on back. Talker, vinyl. 14 in (36 cm). 1970s. Has been re-issued ($50) Scarcity: 6. $50-$200.

Inazuman with pull string talking voice box on back. Talker, vinyl. 15 in (38 cm). 1970s. Has been re-issued ($50). Scarcity: 6. $75-$250.

Kamen Rider Amazon with pull string talking voice box on back. Talker, vinyl. 14 in (36 cm). 1970s. Scarcity: 6. $75-$250.

Kamen Rider V3 with talking voice box and separate belt. Talker, vinyl. 14 in (36 cm). 1970s. Scarcity: 8. $125-$400.

Kamen Rider X with talking voice box and separate belt. Talker, vinyl. 14 in (36 cm). 1970s. Scarcity: 8. $125-$400.

Kikaida with push button talking voice box on back. Talker, vinyl. 16.5 in (42 cm). 1970s. Has been re-issued ($50). Also produced with external stand alone voice box (+50%). Scarcity: 9. $400-$1,000.

Lion Maru with pull string talking voice box on back. Talker, vinyl. 13 in (33 cm). 1970s. Scarcity: 7. $100-$350.

Mach Baron with pull string talking voice box on back. Talker, vinyl. 15 in (38 cm). 1970s. Scarcity: 8. $125-$400.

Mazinger-Z with pull string talking voice box on back. Talker, vinyl. 15 in (38 cm). 1970s. Has been re-issued ($50). Scarcity: 7. $100-$350.

Mirrorman with pull string talking voice box on back. Talker, vinyl. 14 in (36 cm). 1970s. Has been re-issued ($50). Scarcity: 7. $100-$350.

Ninja Arashi with pull string talking voice box on back. Talker, vinyl. 15 in (38 cm). 1970s. Scarcity: 8. $125-$400.

Robocon with pull string talking voice box on back. Talker, vinyl. 11 in (28 cm). 1970s. Scarcity: 8. $125-$350.

Robot Poseidon with pull string talking voice box on back. Talker, vinyl. 14 in (36 cm). 1970s. Scarcity: 10. No price found.

Silver Kamen with pull string talking voice box on back. Talker, vinyl. 14 in (36 cm). 1970s. Scarcity: 5. $50-$200.

Tiger Mask with pull string talking voice box on back. Talker, vinyl. 14 in (36 cm). 1970s. Has been re-issued ($50). Scarcity: 7. $100-$350.

Ultra Seven with pull string talking voice box on back. Talker, vinyl. 14 in (36 cm). 1970s. Has been re-issued ($50). Scarcity: 10. $500-$1,500.

Ultraman Ace with pull string talking voice box on back. Talker, vinyl. 15 in (38 cm). 1970s. Scarcity: 6. $75-$300.

Ultraman Leo with pull string talking voice box on back. Talker, vinyl. 14 in (36 cm). 1970s. Scarcity: 7. $100-$350.

Ultraman Taro with pull string talking voice box on back. Talker, vinyl. 15 in (38 cm). 1970s. Scarcity: 6. $75-$250.

Talking Vinyl Figures – Not Pictured

Astroganger with pull string talking voice box on back. Talker, vinyl. 14 in (36 cm). 1970s. Scarcity: 10. No price found.

Barom 1 with pull string talking voice box on back. Talker, vinyl. 14 in (36 cm). 1970s. Has been re-issued ($50). Scarcity: 6. $75-$250.

Denjin Zaboga with pull string talking voice box on back. Talker, vinyl. 15 in (38 cm). 1970s. Scarcity: 9. $250-$600.

Fireman with pull string talking voice box on back. Talker, vinyl. 14 in (36 cm). 1970s. Scarcity: 9. $250-$600.

Vinyl Figures – Windups

Ultraman Series Characters. 1972. Monsters and Ultra Jack Kneeling, Race Car and Fire Truck.

Japanese Characters   135

Gorbagos (Ultraman) with vibration action. Windup, vinyl. 6 in (15 cm). 1972. Scarcity: 7. $50-$150.

Twintail (Ultraman) with vibration action. Windup, vinyl. 6 in (15 cm). 1972. Scarcity: 8. $50-$200.

#3598 **Pretty Booska** with walking action. Windup, vinyl with tin. 6 in (15 cm). 1968. Scarcity: 7. $100-$150.

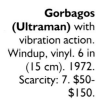

Gudon (Ultraman) with vibration action. Windup, vinyl. 6 in (15 cm). 1972. Scarcity: 7. $50-$150.

Mognezun (Ultraman) with vibration action. Windup, vinyl. 6 in (15 cm). 1972. Scarcity: 7. $50-$150.

Windup Vinyl Figures – Not Pictured

Antlar (Ultraman) with vibration action. Windup, vinyl. 6 in (15 cm). 1972. Scarcity: 7. $50-$150.

Baltan-Seijin (Ultraman) with vibration action. Windup, vinyl. 6 in (15 cm). 1972. Scarcity: 8. $50-$200.

Gomora (Ultraman) with vibration action. Windup, vinyl. 6 in (15 cm). 1972. Scarcity: 7. $50-$150.

Peguira (Ultraman) with vibration action. Windup, vinyl. 6 in (15 cm). 1972. Scarcity: 7. $50-$150.

Ultra Jack kneeling with vibration action. Windup, vinyl. 6 in (15 cm). 1970s. Scarcity: 7. $50-$150 *(shown in group photo of Ultraman Series Characters, page 134)*.

Oba-Q. Obake no Q-Taro ghost with walking action. Windup, vinyl with tin. 6 in (15 cm). 1969. Scarcity: 6. $150-$250.

Oba-Q with Dog. Obake no Q-Taro ghost walks with dog (attached by rod) chasing him. Windup, vinyl with tin. 6 in (15 cm). 1969. Scarcity: 6. $200-$350.

#7509 **Oba-Q with P-Chan**. Obake no Q-Taro ghost walks with sister (attached by rod). Windup, vinyl with tin. 6 in (15 cm). 1969. Scarcity: 6. $200-$350.

Tom & Jerry

Tom & Jerry items, 1975 catalog.

(left) #4239 **Tom On Hand Car** with Tom, ringing bell, and steering action. Battery operated, mystery action, tin with plastic. 10 in (25 cm). 1974-1976. Scarcity: 6. $150-$300. (right) #4240 **Jerry On Hand Car** with Jerry, ringing bell, and steering action. Battery operated, mystery action, tin with plastic. 10 in (25 cm). 1974-1977. Scarcity: 6. $150-$300.

(left) #4209 **Jerry Engineer Locomotive** with moving Jerry, whistle, sound, and lighted lantern. Battery operated, mystery action, plastic with tin. 9.5 in (24 cm). 1974-1976. Scarcity: 4. $75-$150. (right) #4210 **Tom Engineer Locomotive** with moving Tom, whistle, sound, and lighted lantern. Battery operated, mystery action, plastic with tin. 9.5 in (24 cm). 1974-1980. Scarcity: 4. $75-$150.

#4249 **Tom & Jerry Comic Car**. Tom drives car erratically with Jerry trying to stop dynamite explosion. Battery operated, plastic with tin. 12.25 in (31 cm). 1974-1978. Scarcity: 5. $150-$300.

(left) #4251 **Jerry On Buggy** with Jerry and bouncing-spinning action. Windup, mystery action, plastic with tin. 6 in (15 cm). 1974-1976. Scarcity: 4. $75-$125. (right) #4250 **Tom On Jeep** with Tom and bouncing-spinning action. Windup, mystery action, plastic with tin. 6 in (15 cm). 1974-1976. Scarcity: 4. $75-$125.

Tom & Jerry 137

#4256 **Tom & Jerry Fire Engine** with ringing bell and flashing light. Battery operated, mystery action, tin with plastic. 16 in (41 cm). 1974-1975. Scarcity: 5. $125-$225.

#4363 **Tom & Jerry Hot Rod Car** with engine noise, Tom driving and Jerry holding dynamite. Battery operated, mystery action, tin and plastic. 10.5 in (27 cm). 1975-1976. Scarcity: 6. $200-$400.

#4371 **Tom & Jerry Comic Plane** with Tom piloting and Jerry holding on at rudder. Plane does figure 8 while propeller revolves. Battery operated, mystery action, plastic. 10 in (25 cm). 1975-1976. Scarcity: 8. $300-$500.

#4260 **Tom & Jerry Choo Choo**. Tom the engineer plays cymbals and Jerry rides and shakes. Battery operated, mystery action, tin and plastic. 10.5 in (27 cm). 1974-1976. Scarcity: 4. $125-$250.

Art. No. 4262 Jerry on Buggy   Art. No. 4261 Tom on Buggy

(left) #4262 **Jerry On Red Buggy** with Jerry, sound, and bouncing-spinning action. Battery operated, mystery action, plastic with tin. 10.75 in (27 cm). 1974-1977. Scarcity: 5. $125-$250. (right) #4261 **Tom On Green Buggy** with Tom, sound, and bouncing-spinning action. Battery operated, mystery action, plastic with tin. 10.75 in (27 cm). 1974-1977. Scarcity: 5. $125-$250.

138 Character Toys

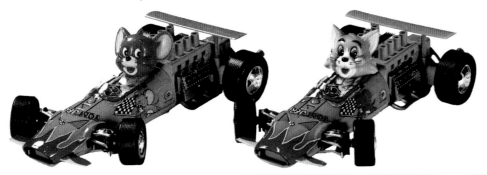

#4372 **Tom & Jerry Race Car** with sound and Tom or Jerry driver. Battery operated, tin and plastic. 11.5 in (29 cm). 1975-1976. Scarcity: 5. $175-$350.

#4374 **Tom & Jerry Patrol Car**. Chevrolet Camaro with Tom & Jerry and flashing light. Battery operated, mystery action, tin with plastic. 11 in (28 cm). 1975-1976. Scarcity: 5. $175-$350.

#4385 **Tom & Jerry Trolley** with bell, flashing light, and moving poles. Tom moves while pulling bell. Battery operated, mystery action, plastic and tin. 11 in (28 cm). 1976. Scarcity: 6. $100-$200.

Art. No. 4485   Art No. 4547

(left) #4485 **Tom & Jerry Lunchbox**. Yellow lunchbox with Tom & Jerry characters. Metal. 10.25 in (26 cm). 1977-1978. Scarcity: 8. $200-$300. (right) #4547 **Droopy Lunchbox**. Red lunchbox with Droopy Dog characters. Metal. 8.25 in (21 cm). 1977-1978. Scarcity: 8. $150-$250.

#4376 **Tom & Jerry Helicopter** with turning blades and sound. Battery operated, non-fall, tin and plastic. 9.5 in (24 cm). 1975-1976. Scarcity: 6. $200-$400.

#4641 **Tom & Jerry Chasing**. Jerry trying to box Tom while attached by wire as Tom moves about and stands. Battery operated, plush and vinyl. 15 in (38 cm). 1978. Scarcity: 5. $100-$200.

Tom & Jerry 139

#4763 **Tom Telephone**. Play rotary telephone with turning dial and ringing bell. Plastic. 7 in (18 cm). 1980. Scarcity: 5. $25-$50.

#7767 **Tom & Jerry Dump Trucks** with ringing bell and dump action. Windup, plastic. 7 in (18 cm). 1976-1980. Scarcity: 4. $40-$75.

#4845 **Tom & Jerry on Donkey Wagon**. Tom & Jerry riding in donkey cart. Windup, mystery action, tin and plastic. 6 in (15 cm). 1981. Scarcity: 5. $30-$60.

#7768 **Tom & Jerry Fire Engines** with ringing bell and moving ladder. Windup, plastic. 7 in (18 cm). 1976-1980. Scarcity: 4. $40-$75.

#7746 **Tom & Jerry Race Cars** with engine sound. Friction, plastic. 7 in (18 cm). 1976-1980. Scarcity: 4. $40-$75.

#7900 **Tom & Jerry Alarm Clock**. Windup, plastic. 4.25 in (11 cm). 1978. Scarcity: 5. $25-$50.

Other Character Toys

(left to right) **Gekko Kamen Rocket Racer**. **Batman Car** with vinyl Batman driver and sound. Friction, tin. 12.75 in (32 cm). 1960s. Scarcity: 9. $1,200-$2,000. **Oba-Q Rocket Racer**.

#2658 **Walther P.P.K Deluxe Set**. James Bond inspired Secret Agent gun with holster, silencer, radio, and papers. Plastic. 1968. Scarcity: 5. $40-$75.

#3622 **Tarzan on Elephant** with Tarzan, monkey, and walking action. Windup, vinyl with tin. 6.25 in (16 cm). 1968-1970. Scarcity: 5. $50-$100.

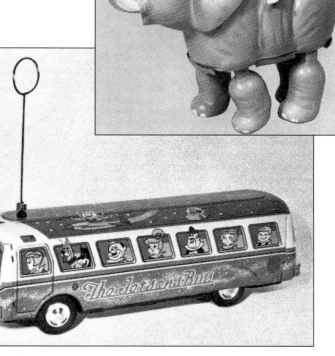

#3371 **Jetsons Bus** with antenna and lithographed Jetsons figures in windows. Friction, tin. 14 in (36 cm). 1965-1966. Scarcity: 10. $4,000-$6,000.

Other Character Toys 141

Other – Not Pictured

Topo Gigio on 3-Wheel Scooter. Vinyl Topo Gigio on flat 3-wheel scooter. Windup, tin and vinyl. 7.75 in (20 cm). 1960s. Scarcity: 6. $125-$250.

#2680 Falcon Super Deluxe Walther PPK & Beretta. James Bond inspired Secret Agent Walthers and Beretta with holster, silencer, radio, and papers. Plastic. 1968-1972. Scarcity: 6. $50-$100.

#4340 **Karate Champion** puppet with arms that hit quickly when activated internally. Vinyl. 10 in (25 cm). 1975-1976. Scarcity: 4. $20-$40.

#4446 **Yogi Bear on Comic Car** with musical voice sound. Friction, plastic. 7.5 in (19 cm). 1976. Scarcity: 6. $50-$100.

#4340A **Bruce Lee Karate** puppet with arms that hit quickly when activated internally. Vinyl. 10 in (25 cm). 1976. Scarcity: 6. $40-$75.

#4456 **Comic Dog Helicopter with Missiles.** Dastardly and Mutley-like dog on copter with shooting missiles. Windup, plastic. 7.5 in (19 cm). 1976. Scarcity: 5. $30-$50.

#4524 **Woody Woodpecker Boat** with removable outboard motor. Windup, plastic. 7.5 in (19 cm). 1977. Scarcity: 5. $30-$50.

Circus and Clowns

Many toys have been produced depicting animals at the circus and the always entertaining clowns. This section contains both types of toys.

Circus Animals

#1970 **Hoop Spinning Circus Seal**. Plush covered seal spinning hoop on nose. Windup, plush. 5.5 in (14 cm). 1960. Scarcity: 3. $40-$75.

Circus Bear twirling umbrella with attached balls. Windup, celluloid. 7 in (18 cm). 1940s. Scarcity: 7. $150-$300.

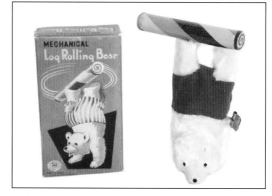

#3086 **Happy Time Circus Animals (Bear, Seal, Elephant)**. Bear doing hand stand with spinning log on its foot, Elephant spinning ball or seal spinning hoop. Windup, plush and tin. 6.5 in (17 cm). 1960-1962. Scarcity: 4. $50-$100.

Circus Elephant with raising head and twirling tail. Windup, plush and tin. 6 in (15 cm). 1940s. Scarcity: 3. $75-$125.

#1881 **Circus Seal**. Walking seal with spinning ball. Windup, plush and tin. 8 in (20 cm). 1958. Scarcity: 5. $60-$100.

(left) #4661 **Tricycler Bear** with plush covered bear and ringing bell. Windup, plush and plastic with tin. 7 in (18 cm). 1978-1980s. Scarcity: 3. $15-$30. (right) #4660 **Tricycler Monkey** with plush covered monkey and ringing bell. Windup, plush and plastic with tin. 7 in (18 cm). 1978-1980s. Scarcity: 3. $15-$30.

Clowns 143

Circus Animals – Not Pictured

#1070 Rollo The Monk. Celluloid monkey rolling in tin wire hoop. Windup, tin and celluloid. 5 in (13 cm). 1940s. Scarcity: 6. $150-$300.

#1884 Beast and Clown. Lion chasing clown up striped pole. Windup, tin. 16 in (41 cm). 1958. Scarcity: 7. $150-$300.

#1936 Sparky the Seal. Sparky juggles celluloid ball that floats on air. R/C battery operated, plush. 9.75 in (25 cm). 1959. Scarcity: 3. $50-$100.

#3083 Circus Elephant spins ball on trunk while standing on hind legs. Windup, plush and tin. 8 in (20 cm). 1960-1962. Scarcity: 4. $75-$125.

Clowns

#3478 **Pierrot-Monkey Cycle.** Clown with monkey waving hat on hand car with sound and red nose light. Battery operated, mystery action, tin with vinyl head. 10 in (25 cm). 1966-1968. Scarcity: 6. $225-$450.

Ninkimono with spinning bar of balls and bells balanced on nose. Windup, celluloid and wood. 13 in (33 cm). 1930s. Scarcity: 7. $175-$350.

#3553 **Clown and Lion** with lion chasing a teasing clown up the tree. Battery operated, tin and vinyl. 13.5 in (34 cm). 1967-1970. Scarcity: 6. $175-$350.

#1734 **Circus Hoop.** Clown rides unicycle in hoop. Windup, tin. 6.5 in (17 cm). 1957-1958. Scarcity: 7. $250-$450.

144 Circus and Clowns

#1748 **Jolly Clown** with cane, folding arms and walking action. Windup, tin with cloth. 7 in (18 cm). 1957-1958. Scarcity: 7. *Les Fish collection*. $125-$225.

#3219 **Tric Cycling Clown**. Clown with blinking lights in each hand, peddles on unicycle in square pattern. Battery operated, tin with plastic. 5.75 in (15 cm). 1962-1964. Scarcity: 7. *Courtesy of Smith House Toys*. $250-$450.

#1748B **Jolly Clown** with tambourine, bell, folding arms, and walking action. Windup, tin with cloth. 7 in (18 cm). 1957-1958. Scarcity: 10. *Les Fish collection*. $400-$650.

#3479 **Circus Parade Car w/Horn** with clown driver, shaking and sound. Battery operated, mystery action, tin. 12.75 in (32 cm). 1966-1967. Scarcity: 5. $200-$400.

#3317 **Circus Clown** hangs on wire stand and does body flips. Can be positioned multiple ways. Windup, tin and cloth. 7 in (18 cm). 1964-1968. Scarcity: 6. $125-$250.

Clowns 145

#3480 **Circus Fire Engine w/Bell** with clown driver, comic car action, and sound. Battery operated, tin. 11 in (28 cm). 1966-1969. Scarcity: 6. $125-$250.

#3855 **Hop-Up Clown Car** with clown that hops up while driving. Battery operated, mystery action, tin and plastic. 10.5 in (27 cm). 1970-1973. Scarcity: 4. $100-$175.

Clowns – Not Pictured

Fancy Dan the Juggling Man. Juggler twirling hat on nose. Windup, celluloid. 6 in (15 cm). 1950s. Scarcity: 6. $200-$400.

Showa no Kodomo (Child of the 1930s). Windup, celluloid. 9.5 in (24 cm). 1930s. Scarcity: 10.

#0366 **Topsy Turvy Tom**. Painted car with driver and rollover action. Windup, tin. 8.5 in (22 cm). 1930s. Scarcity: 8. $400-$600.

#1104 **Baby Tumbling Pierrot** with long rotating arms that cause clown to tumble. Windup, celluloid. 5 in (13 cm). 1940s. Scarcity: 4. $75-$150.

#1105 **Pierrot Tricycle**. Celluloid clown standing on big wheel tin tricycle with bell. Windup, celluloid and tin. 4.75 in (12 cm). 1940s. Scarcity: 5. $150-$300.

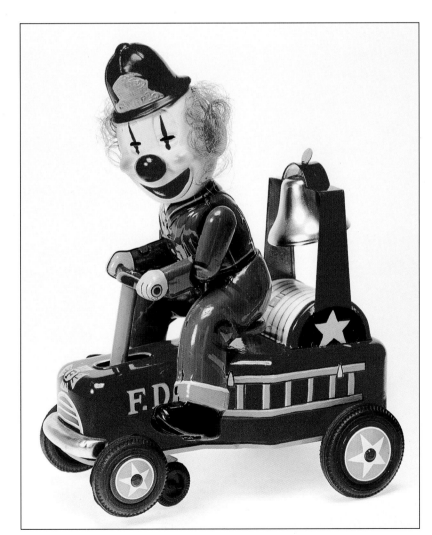

#3766 **Clown Hand Car** with tin fire clown driver on fire engine hand car with bell. Battery operated, mystery action, tin with vinyl. 10 in (25 cm). 1969-1972. Scarcity: 5. $150-$275.

Farm Tractors

Farm Tractor Set with tin driver, wagon, sickle bar, and rake. Friction, tin. 1950s. Scarcity: 6. $200-$350.

(bottom) #1484 **Tractor** with tin driver. Friction, tin. 5.5 in (14 cm). 1955-1960. Scarcity: 5. $75-$150. (top) #1661 **Tractor No. 100** with tin driver. Battery operated, tin. 8 in (20 cm). 1956-1960. Scarcity: 5. $125-$250.

Farm Tractors – Not Pictured

#1695 **Tractor Set No.2** with tin driver, spreader, trailer, and sickle bar. Friction, tin. 7 in (18 cm). 1957-1960. Scarcity: 6. $100-$200.

#1708 **Hi Powered Tractor No.102 with Conveyor Car** includes sand bucket conveyor. Battery operated, tin. 20.5 in (52 cm). 1957-1960. Scarcity: 5. $125-$250.

#1740 **Tractor with Visible Engine** with tin driver, steering and visible engine. R/C battery operated, tin. 8 in (20 cm). 1958. Scarcity: 4. $75-$150.

#1826 **Tractor with Trailer Piston Action**. Tractor No.110 with driver and visible moving pistons pulling 2-wheeled trailer. Battery operated, tin. 14.75 in (37 cm). 1958-1960. Scarcity: 6. $100-$200.

#1997 **Farm Tractor w/ Trailer**. 5.5-inch tractor with driver (Cat #1484) pulling trailer. Friction, tin. 11 in (28 cm). 1960-1967. Scarcity: 5. $100-$175.

#3373 **Farmighty Largest Tractor** moves back and forth with engine sound and flashing light. Battery operated, tin with vinyl. 10.5 in (27 cm). 1965-1969. Scarcity: 5. $125-$200.

#3578A **Bob's Farm Tractor** with moving vinyl driver, engine sound, and flashing light. Battery operated, mystery action, tin with vinyl. 10.5 in (27 cm). 1967-1970. Scarcity: 4. $100-$175.

#3578B **Bob's Farm Tractor** with vinyl driver, flashing pistons, and engine sound. Battery operated, mystery action, tin with plastic. 10.5 in (27 cm). 1972-1973. Scarcity: 4. $75-$150.

Games

Art. No. 2077
Rifle Target Game

Art. No. 2078
Junior Gun Set

Art. No. 2079
Junior Gun Set

(top) #2077 **Rifle Target Game**. Cork firing rifle with scope and plastic targets. Plastic. 21 in (53 cm). 1980s. Scarcity: 4. $20-$40. (middle) #2078 **Junior Gun Set**. Cork firing rifle with plastic targets. Plastic. 19.5 in (50 cm). 1980s. Scarcity: 4. $20-$40. (bottom) #2079 **Junior Gun Set**. Cork gun with plastic targets. Tin and plastic. 9 in (23 cm). 1980s. Scarcity: 4. $20-$40.

(top) #3481 **Big Mouth Gorilla Target Game**. Mighty Kong Big Mouth Target Game with ball floating on air and dart gun with three darts. Battery operated, tin and plastic. 13 in (33 cm). 1967-1968. Sold by Marx. Scarcity: 9. $300-$500. (bottom) #3570 **Big Mouth Chief Target Game**. Indian target with ball floating on air and dart pistol with three darts. Battery operated, tin with plastic. 13 in (33 cm). 1967-1968. Scarcity: 8. $150-$300.

#3572 **Family Racing "3 Lane."** 3 lane and car slot car set with power pack, sectional track, guard rails, and Jaguar, Lotus Elan and Prince R-380 cars. Battery operated, plastic. 1967-1969. Scarcity: 7. $50-$100.

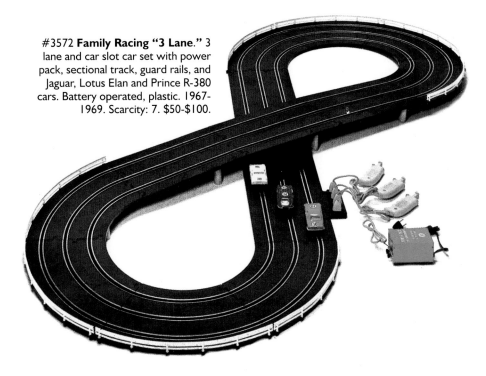

148   Games

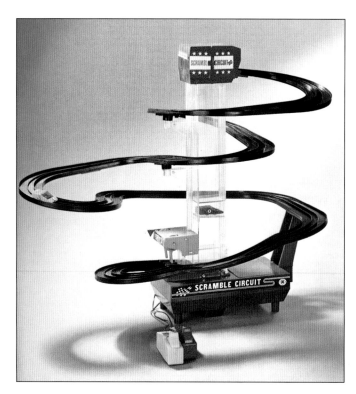

#4550 **Scramble Circuit** with 2-lane raceway and air lift to take race cars to top of track for racing. Battery operated, plastic. 21 in (53 cm). 1977-1980s. Scarcity: 3. $25-$75.

#4280 **Challenge Golf Game**. Lay out your own course on green mat then position remote control golfers to actually play the course. Cloth and plastic. 69 in (175 cm). 1975. Scarcity: 5. $50-$75.

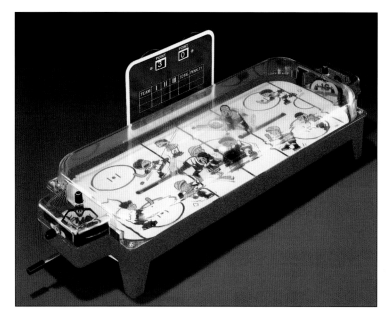

#4549 **Ice Hockey Game**. Enclosed hockey game with player control over one player and goalie. Plastic. 20 in (51 cm). 1978-1979. Scarcity: 3. $25-$45.

#4880 **Karate Man Game**. Players attempt to chop bricks. Plastic. 1982. Scarcity: 6. $15-$25.

Games 149

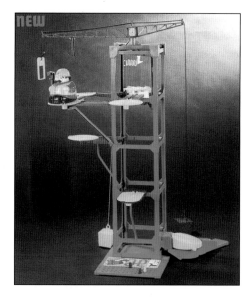

#4613 **Sleepwalker Game**. Construction worker "Sleepwalkin' Sam" is manipulated through the tower structure. Windup, plastic. 22 in (56 cm). 1978-1979. Scarcity: 4. $20-$40.

#4659 **Fishing Game**. Magnetic fishing game with 20 fish and 4 fishing rods with magnet line. Windup, plastic. 12 in (30 cm). 1978-1979. Scarcity: 4. $20-$40.

#4912 **Earthquake Game**. Players attempt to stack blocks to build earthquake proof towers. Plastic. 1982. Scarcity: 6. $20-$35.

#7906 **Worm Wrestling Game**. Players try to push magnetic worms out of the ring. Plastic. 9.5 in (24 cm). 1978. Designed by Marvin Glass of Chicago, Illinois. Scarcity: 5. $20-$40.

Games – Not Pictured

#1837 **Shooting Set**. Dart shooting bow and arrow set with tin lion face target board. Tin. 6.5 in (17 cm). 1958-1960. Scarcity: 3. $40-$75.

House Play Toys

This broad category often catered to little girls but the toys had a broad appeal to the younger child. Miniature household appliances and cookware, dolls, baby toys, and telephones are included here.

Appliances and Cookware

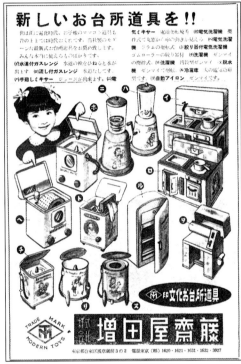

Trade journal ad, April 1956.

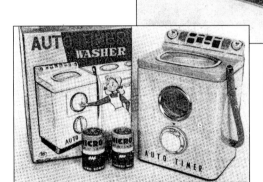

#2167 **Kiddies Typewriter Typet**. Play typewriter with working print wheel. Tin and plastic. 11.25 in (29 cm). 1962-1968. Scarcity: 4. $25-$50.

#1803 **Auto Timer Washer** with spinning action, timer, and rubber hose. Battery operated, tin. 7.5 in (19 cm). 1957-1961. Scarcity: 4. $50-$100.

#1621b **Automatic Washing Machine No. 2** with lid, agitator, side basket, and rubber hose. Two children illustrated on front. Windup, tin. 5 in (13 cm). 1960-1967. Scarcity: 5. $50-$100.

#1598 **Washing Machine No.1 (Round)** on legs with lid, agitator, and rubber hose. Windup, tin. 5 in (13 cm). 1956-1960. Scarcity: 5. $50-$100.

Appliances and Cookware   151

#1635 **Refrigerator** with ice box and lithographed food on shelves and inside door. Tin. 7 in (18 cm). 1956-1960. Scarcity: 5. $75-$150.

#1664 **Little Miss Automatic Ironer** with fabric swatch for ironing and on/off switch. Windup, tin. 4.5 in (11 cm). 1956-1960. Scarcity: 5. $65-$125.

#1641 **Spin Dryer**. Round automatic spin dryer with legs and rubber drain hose. Windup, tin. 5 in (13 cm). 1956-1960. Scarcity: 6. $60-$120.

#1710 **Kiddy Television** with rotating picture wheel and musical sound. Windup, tin. 5.5 in (14 cm). 1957-1960. Scarcity: 6. $125-$250.

#1646 **Washing Machine No. 10** with hose, agitator, wringer, and front on-off switch. Battery operated, tin. 6 in (15 cm). 1956-1962. Scarcity: 5. $75-$125.

#1723 **Modern Living Set No.2** with miniature refrigerator, stove, TV, blender, and vacuum cleaner. Tin. Box size: 9 in (23 cm). 1957-1960. Scarcity: 9. $150-$300.

House Play Toys

#1752 **Wash-O-Matic Electro Toy**. Miniature washing machine with signal light, front controls, and rubber hose. Battery operated, tin. 6.5 in (17 cm). 1957-1958. Scarcity: 4. $50-$100.

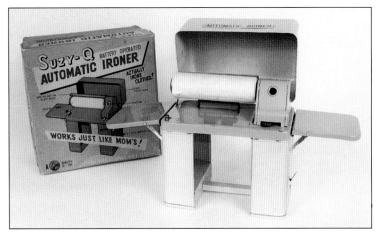

#1779 **Suzy Q Automatic Ironer** works just like Mom's! With fold down side panels and light. Battery operated, tin. 12 in (30 cm). 1957-1958. Scarcity: 4. $75-$150.

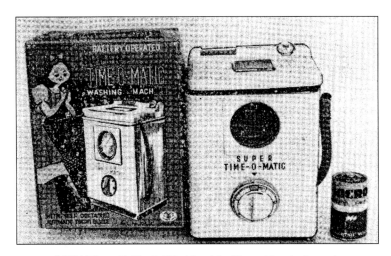

#1804 **Time-O-Matic Washing Machine** with spinning action, timer, and rubber hose. Battery operated, tin. 6 in (15 cm). 1958-1961. Scarcity: 4. $40-$80.

#1847 **Lil Miss Sewing Machine**. Friction, tin. 7 in (18 cm). 1958-1960. Scarcity: 4. $40-$80.

#2377 **Old Fashioned Stove** with opening door, cook top, and utensils. Tin. 7 in (18 cm). 1965-1973. Scarcity: 4. $40-$80.

#2380 **Kitchen Sink** with opening doors, sink, and utensils. Tin. 6 in (15 cm). 1965-1967. Scarcity: 4. $40-$75.

Appliances and Cookware   153

#2383 **Little Miss Electric Stove** with opening doors, cook top, and utensils. Tin. 8.5 in (22 cm). 1965-1970. Scarcity: 4. $40-$80.

#2390 **Lil Miss Kitchen** consisting of washing machine, refrigerator, range, and sink with utensils. Tin. 17.75 in (45 cm). 1965-1967. Scarcity: 5. $85-$150.

#2387 **Automatic Washer**. Laundro-Matic washer with agitator and rubber hose. Windup, tin. 7 in (18 cm). 1965. Scarcity: 5. $40-$80.

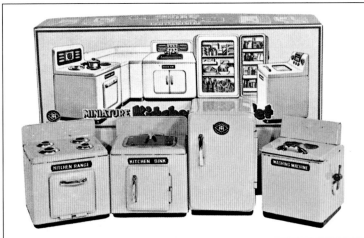

#2396 **Miniature Kitchen 4 Pc. Set** consisting of kitchen range, kitchen sink, washing machine, and refrigerator. Tin. 14 in (36 cm). 1965-1971. Scarcity: 5. $75-$125.

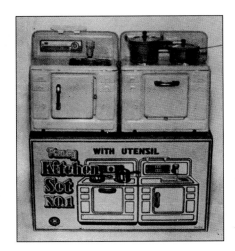

#2388 **Fancy Kitchen Set No.1** consisting of range and sink with utensils. Tin. 10.5 in (27 cm). 1965-1967. Scarcity: 4. $40-$80.

#2389 **Fancy Kitchen Set No.2** consisting of refrigerator, range, and sink with utensils. Tin. 14 in (36 cm). 1965-1970. Scarcity: 4. $65-$100.

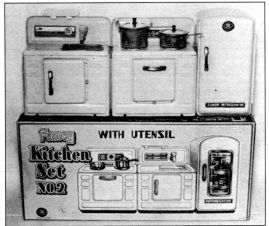

154 *House Play Toys*

#2397 **Miniature Kitchen 3 Pc. Set** consisting of kitchen range, kitchen sink, and refrigerator. Tin. 10.5 in (27 cm). 1965-1972. Scarcity: 3. $50-$80.

#3313 **Super Washing Machine**. Wash-O-Matic with agitator, basket, and hose. Windup, tin with plastic. 8 in (20 cm). 1964. Scarcity: 6. $40-$80.

#2293 **Modern Living Kitchen Set**. Set of stainless steel pots and pans. 1965-1967. Scarcity: 5. $25-$50.

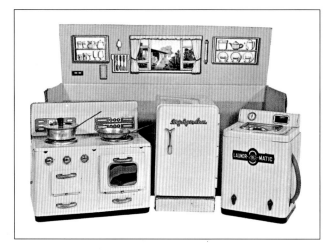

#2399 **Little Lady 3 Units Kitchen Appliance Set** consisting of stove, refrigerator, and washing machine. Tin. 19 in (48 cm). 1965-1968. Scarcity: 5. $125-$225.

#2716 **New Living Kitchen Set No.1**. 10 piece cookware set. Tin. 1968-1973. Scarcity: 3. $20-$30.

#3281 **Junior Automatic Washer** with laundry basket. Battery operated, tin. 7.5 in (19 cm). 1963-1967. Scarcity: 4. $40-$80.

(left) #2720 **Modern Cooking Ware No. 3**. 19 piece cookware set. Tin. 1968-1972. Scarcity: 3. $30-$40. (right) #2718 **New Living Kitchen Set No.3**. 18 piece cookware set. Tin. 1968-1972. Scarcity: 3. $30-$40.

Baby Toys   155

Appliances and Cookware – Not Pictured

#1621a Automatic Washing Machine No. 2 with lid, agitator, and rubber hose. Single child illustrated on front. Windup, tin. 5 in (13 cm). 1956-1960. Scarcity: 5. $50-$100.

#1663 Junior Automatic Washer with light, spinning drum, hose, and controls on front. Battery operated, tin with plastic. 7 in (18 cm). 1956-1960. Scarcity: 5. $50-$100.

#1703 Baby Television with Music with music, picture, and battery operated light. Windup, tin. 6.5 in (17 cm). 1957-1960. Scarcity: 7. $125-$200.

#1709 Refrigerator No.10 with Light with ice box and lithographed food on shelves and inside door plus light. Battery operated, tin. 8.5 in (22 cm). 1957-1960. Scarcity: 4. $60-$125.

#1719 Modern Living Set No.1 with miniature refrigerator, stove, and sink. Tin. 10.75 in (27 cm). 1957-1960. Scarcity: 5. $100-$200.

#1745 Toaster. Little Miss Housekeeping miniature toaster. Battery operated, tin. 6 in (15 cm). 1957-1958. Scarcity: 4. $50-$100.

#1768 Washing Machine No. 3. Little Miss Housekeeper washing machine with rubber hose and see-through plastic top. Windup, tin and plastic. 4.5 in (11 cm). 1957-1958. Scarcity: 5. $40-$80.

#1769 Hand Blender. Miniature blender with plastic top and hand crank. Tin and plastic. 7 in (18 cm). 1957-1958. Scarcity: 4. $40-$80.

#1780 Mangle Ironer with fold down side panels and push button friction inertia motor. Friction, tin. 9.5 in (24 cm). 1957-1958. Scarcity: 6. $75-$150.

#1806 Quizzer with Arithmetic. Arithmetic quizzing machine. Tin. 5.5 in (14 cm). 1958-1961. Scarcity: 3. $20-$30.

#1855 Lil Miss Mixer with mixing tool. Friction, tin and plastic. 6.5 in (17 cm). 1958-1960. Scarcity: 2. $20-$40.

#2379 Kitchen Range with opening double doors, cook top, and utensils. Tin. 6 in (15 cm). 1965-1967. Scarcity: 3. $35-$60.

#2381 Refrigerator With Fruits. See-through opening door refrigerator with plastic fruit. Tin with plastic. 6.5 in (17 cm). 1965. Scarcity: 6. $50-$95.

#2384 Refrigerator With Freezer. Tin. 8 in (20 cm). 1965. Scarcity: 6. $50-$75.

#2717 New Living Kitchen Set No.2. 16 piece cookware set. Tin. 1968-1972. Scarcity: 3. $25-$35.

#3171 Washing and Drying Machine. Washing machine with timer, rubber hose, plastic lid, and spin dry drum. Battery operated, tin. 7 in (18 cm). 1961-1962. Scarcity: 4. $30-$60.

#3271 Helicopter Roller. Animals helicopter shaped push toy. Tin with plastic. 21 in (53 cm). 1963-1967. Scarcity: 3. $20-$40.

#3389 Butterfly Roller. Butterfly push toy with flapping wings. Tin. 22 in (56 cm). 1965-1967. Scarcity: 3. $20-$40.

Baby Toys – Not Pictured

#1692 Baby Roller No. 2. Wood handled push toy with tin wheels and ball in cage. Tin and wood. 25 in (64 cm). 1957-1960. Scarcity: 2. $20-$40.

#1712 Baby Roller No. 3. Wood handled push toy with tin wheels and ball. Tin and wood. 25 in (64 cm). 1957-1960. Scarcity: 2. $20-$40.

#1721 Musical Roller. Push toy, wood and tin. 27 in (69 cm). 1957. Scarcity: 2. $25-$50.

#1810 Butterfly Roller. Butterfly with flapping wings on wheeled push toy. Tin and wood. 23 in (58 cm). 1958-1961. Scarcity: 3. $25-$50.

#1953 No. 12 Baby Roller with zoo animal graphics. Tin and wood or vinyl. 22 in (56 cm). 1960-1962. Scarcity: 2. $20-$40.

Baby Toys

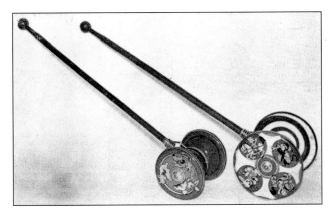

(left) **#1879 Baby Roller No.11**, also shown in adjacent photo. (right) **#1660 Baby Roller No.1**. Wood or vinyl handled push toy with tin wheels and ball and kiddy horse graphics. Tin and wood. 23 in (58 cm). 1955-1970. Scarcity: 2. $20-$40.

#1879 Baby Roller No.11. Plastic handled push toy. Tin with vinyl. 17 in (43 cm). 1958-1976. Scarcity: 2. $15-$30.

156 House Play Toys

Dolls

#3010/4073 **Crawling Baby**. Baby crawls on hands and knees. Windup, vinyl, tin and cloth. 5 in (13 cm). 1960-1976. Became #4073 in 1972 catalog. Scarcity: 2. $20-$40.

#1388 **Walking Doll** dressed with painted sleep eyes and walking action. Windup, celluloid and tin. 14 in (36 cm). 1952-1960. Scarcity: 5. $250-$400.

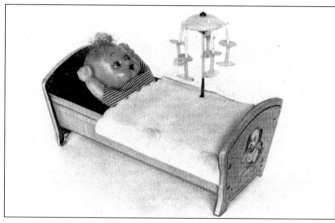

#3109 **Happy Time Baby**. Vinyl baby doll in bed with spinning toy. Windup, tin, vinyl and cloth. 8.5 in (22 cm). 1960-1964. Scarcity: 4. $40-$75.

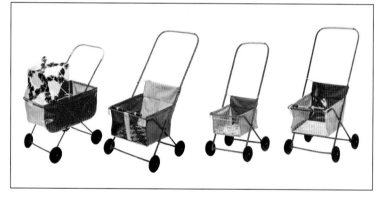

(left to right) #2033 **Baby Carriage** with canopy. Vinyl and metal. 23.25 in (59 cm). 1962-1973. Scarcity: 3. $20-$30. #2295 **Baby Carriage** stroller. Vinyl and metal. 27 in (69 cm). 1965-1980s. Scarcity: 3. $20-$30. #2060 **Baby Carriage** stroller. Vinyl and metal. 22 in (56 cm). 1962-1980s. Scarcity: 2. $15-$25. #2229 **Baby Carriage** stroller. Vinyl and metal. 21 in (53 cm). 1965-1980s. Scarcity: 3. $15-$25.

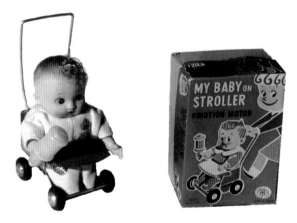

#3114 **My Baby on Stroller**. Baby in small stroller with baby rattle. Friction, tin and vinyl. 5 in (13 cm). 1961-1964. Scarcity: 3. $30-$60.

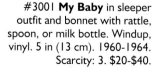

#3001 **My Baby** in sleeper outfit and bonnet with rattle, spoon, or milk bottle. Windup, vinyl. 5 in (13 cm). 1960-1964. Scarcity: 3. $20-$40.

#3337 **My Crawling Baby** moves forward on hands and knees. Striped outfit. Windup, vinyl, cloth and tin. 5 in (13 cm). 1964-1967. Scarcity: 3. $25-$45.

Dolls 157

#3931 **Crawling Baby** crawls forward, stops and cries before repeating action. Battery operated, vinyl. 9.5 in (24 cm). 1971-1977. Scarcity: 2. $20-$40.

#7733 **Miniature Doll**. Pulling string on back of doll causes her to play. Vinyl. 6 in (15 cm). 1976-1979. Designed by Marvin Glass of Chicago, Illinois. Scarcity: 2. $20-$30.

#3360 **Humorous Baby** with bottle moves forward while shaking bottle. Striped outfit. Windup, vinyl, cloth and tin. 6 in (15 cm). 1965-1967. Scarcity: 3. $25-$45.

#7770 **Miniature Doll on Swing**. Pulling string on back of doll causes her to play. Vinyl with plastic. 11 in (28 cm). 1976-1979. Designed by Marvin Glass of Chicago, Illinois. Scarcity: 3. $20-$40.

#7771 **Miniature Doll on Rocking Horse**. Pulling string on back of doll causes her to play. Vinyl. 7 in (18 cm). 1976-1979. Designed by Marvin Glass of Chicago, Illinois. Scarcity: 3. $20-$40.

#3736 **Make-up Suzy** applies make-up to her face. Windup, vinyl. 5.25 in (13 cm). 1969-1971. Scarcity: 5. $40-$75.

#3962 **Dancing Doll**. Three different dolls representing kids dressed as cowboys and Indians vibrate and move in circle. Windup, vinyl. 5.25 in (13 cm). 1971-1973. Scarcity: 3. $15-$30.

#7731 **Miniature Doll in Rocking Chair**. Pulling string on back of doll causes her to play. Vinyl. 7 in (18 cm). 1976-1979. Designed by Marvin Glass of Chicago, Illinois. Scarcity: 3. $20-$40.

Dolls – Not Pictured

#2029 **Baby Carriage** with canopy. Vinyl and metal. 18.5 in (47 cm). 1962-1972. Scarcity: 3. $20-$30.

#2059 **Baby Carriage** stroller. Vinyl and metal. 15 in (38 cm). 1962-1969. Scarcity: 3. $15-$25.

#3024 **My Toddler** with baby rattle in circular walker. Windup, vinyl, cloth and tin. 6 in (15 cm). 1960. Scarcity: 3. $30-$50.

#3096 **Toddling Baby** with baby rattle in circular walker. Windup, cloth, tin and vinyl. 8 in (20 cm). 1960-1962. Scarcity: 3. $35-$55.

#7730 **Miniature Doll in High Chair**. Pulling string on back of doll causes her to play. Vinyl. 7 in (18 cm). 1976-1979. Designed by Marvin Glass of Chicago, Illinois. Scarcity: 3. $20-$40.

#7732 **Miniature Doll in Cradle**. Pulling string on back of doll causes her to play. Vinyl. 6 in (15 cm). 1976-1979. Designed by Marvin Glass of Chicago, Illinois. Scarcity: 3. $20-$40.

Miniatures

#2186 **Super Market**. Supermarket box with miniature foods. Paper with plastic. 17.25 in (44 cm). 1965-1969. Scarcity: 6. $30-$55.

#2515 **Food Shop "Grocery."** Grocery Store with groceries. Paper with plastic. 9.75 in (25 cm). 1966-1975. Scarcity: 6. $20-$40; #2516 **Food Shop "Vegetable."** Food shop with vegetables. Paper with plastic. 9.75 in (25 cm). 1966-1975. Scarcity: 6. $20-$40; #2517 **Food Shop "Fruits."** Food shop with fruits. Paper with plastic. 9.75 in (25 cm). 1966-1975. Scarcity: 6. $20-$40; #2518 **Food Shop "Meat Shop."** Food shop with meats. Paper with plastic. 9.75 in (25 cm). 1966-1975. Scarcity: 6. $20-$40

#2947 **Bonmart Ice Cream Shop**. Ice Cream shop play set with Coca-la advertising (in Japanese). Paper and plastic. 11.5 in (29 cm). 1970-1971. Scarcity: 4. $50-$100.

Miniatures – Not Pictured

#2096 **Refrigerator Set**. Refrigerator-like box with miniature contents. Paper with plastic. 10.75 in (27 cm). 1962-1969. Scarcity: 5. $25-$50.

Telephones

#1801 **Baby Phone**. Miniature rotary dial telephone. Tin. 4.5 in (11 cm). 1958-1962. Scarcity: 4. *Courtesy of Barbara Moran.* $70-$125.

(top) #3039 **Handy Talkie Set**. Walkie-Talkie set with 32 foot wire cord attached. Battery operated, plastic. 14 in (36 cm). 1959-1976. Scarcity: 2. $20-$40.

(bottom) #3039B **Walkie Talkie**. Military version of Walkie-Talkie set with 32 foot wire cord attached. Battery operated, plastic. 14 in (36 cm). 1966-1976. Scarcity: 2. $20-$40.

Telephones 159

(left) #3170 **Ring Ring Phone**. Rotary play telephone with ringing bell when cord is pulled. Tin. 7 in (18 cm). 1961-1964. Scarcity: 5. $40-$75. (right) #3214 **Baby Ring Ring Phone**. Rotary play telephone with ringing bell when cord is pulled. Tin. 6 in (15 cm). 1962-1964. Scarcity: 5. $40-$75.

#3383 **Pla-Lovely Phone** with ringing bell when dial is turned. Battery operated, plastic. 8.5 in (22 cm). 1965-1967. Scarcity: 3. $20-$30.

#3214 **Baby Ring Ring Phone**. Rotary play telephone with ringing bell when cord is pulled. Tin. 6 in (15 cm). 1962-1964. Scarcity: 5. $40-$75.

Art. No. 3519   New Junior Phone

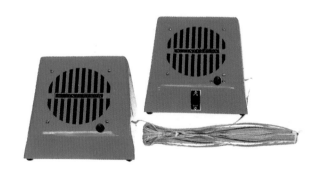

Art. No. 3718   Hi-Inter Phone

(left) #3519 **New Junior Phone**. Two phone set with connecting line cord and dial ringer. Battery operated, plastic. 1967-1976. Scarcity: 5. $30-$50. (right) #3718 **Hi-Interphone** working intercom. Battery operated, plastic. 14.5 in (37 cm). 1969-1973. Scarcity: 3. $30-$50.

Telephones – Not Pictured

#1764 **Junior Phone** with two hand phones with long wire between them. Battery operated, tin. 1957-1958. Scarcity: 5. $40-$75.

#1843 **Junior Phone -2L** twin play phone set. Battery operated, plastic. 8 in (20 cm). 1958-1960. Scarcity: 4. $25-$50.

#1924 **Communication Set**. Dual Morse code transmitters and receivers. Battery operated, tin with plastic. 8 in (20 cm). 1959. Scarcity: 5. $40-$75.

#1928 **Ring Ring Phone**. Miniature rotary dial telephone with girl face illustrated on center of dial. Phone rings when cord is pulled. Tin. 7 in (18 cm). 1959. Scarcity: 5. $75-$125.

#3220 **Junior Phone**. Working play phone with ringing bell. Battery operated, plastic. 8 in (20 cm). 1962-1967. Scarcity: 4. $20-$40.

#3367 **Buzz-a-Phone** with push button buzzer. Battery operated, plastic. 12.5 in (32 cm). 1965-1966. Scarcity: 3. $20-$40.

#3384 **Pla-Play a Ring Phone** rings when dial is pushed. Battery operated, plastic. 8 in (20 cm). 1965-1967. Scarcity: 3. $20-$30.

#3366 **Bellite Phone** with bell ringer and connecting cord. Battery operated, plastic. 8 in (20 cm). 1965-1970. Scarcity: 3. $20-$40.

Military

This section features toys depicting military action. Armored cars, cannons, guns, jeeps, tanks, and trucks are all included in this listing. Soldiers are included in the People category.

Armored Cars – Not Pictured

#0300 **Firing Car**. Sparking wheeled vehicle with gun and painted camouflage. Friction, tin. 3 in (8 cm). 1930s. Scarcity: 7. $125-$200.

Armored Cars

Armored Car. Painted camouflage truck with shovels and large moveable cannon. Friction, tin. 5.5 in (14 cm). 1950. Scarcity: 4. $75-$150.

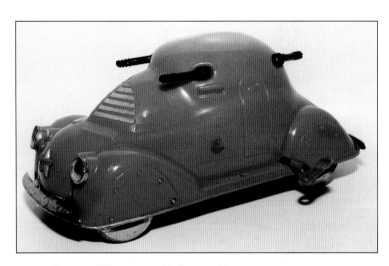

Armored Motorcar with four machine guns. Windup, celluloid. 8 in (20 cm). 1930s. Scarcity: 10. *Bill & Stevie Weart Collection*. $1,000-$1,500.

Cannons & Guns

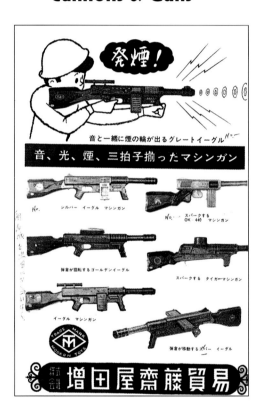

Trade journal ad, March 1962.

Cannon. Camouflaged cannon in red, yellow, and blue. Spring loaded, tin. 8 in (20 cm). 1950s. Scarcity: 4. $75-$150.

Cannons & Guns   161

Flash Toy with sparking. Friction, tin. 4 in (10 cm). 1950s. Scarcity: 3. $20-$30.

#1907 **Condor Gun**. Shooting gun with ten plastic bullets. Tin, plastic. 15.5 in (39 cm). 1958-1962. Scarcity: 3. $25-$50.

#1988 **Baby Machine Gun OK-10**. "Super" machine gun with sparking. Windup, tin. 10 in (25 cm). 1960. Scarcity: 3. $40-$75.

#3116 **Super Eagle Machine Gun** with horizontal sliding magazine and blinking light. Battery operated, tin. 24 in (61 cm). 1960-1962. Scarcity: 5. $75-$125.

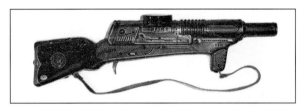

#3009 **Tiger Machine Gun with Sparkling** with tiger head on stock. Friction, tin. 19.5 in (50 cm). 1960-1964. Scarcity: 4. $50-$100.

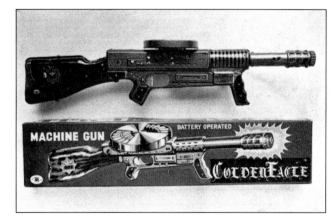

#3135 **Golden Eagle Machine Gun** with circular rotating magazine and blinking light. Battery operated, tin. 24 in (61 cm). 1961-1967. Scarcity: 5. $75-$125.

#3094 **Machine Gun O.K.-220**. OK-220 machine gun. Friction, tin. 14 in (36 cm). 1960-1968. Scarcity: 4. $40-$75.

#3145 **OK-330 Machine Gun**. Friction, tin. 13.5 in (34 cm). 1961-1962. Scarcity: 4. $50-$100.

162 Military

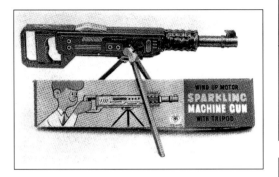

#3148 **Sparkling Machine Gun w/ Tripod** with sparking when fired. Windup, tin. 12.5 in (32 cm). 1961-1962. Scarcity: 5. $75-$150.

#3149 **Machine Gun OK-440 with Spark**. Horse head machine gun with sparking. Friction, tin. 19 in (48 cm). 1961-1964. Scarcity: 4. $75-$125.

#3207 **Great Eagle Smoky Machine Gun**. Eagle 3207 machine gun with, shooting noise, blinking light, and smoking. Battery operated, tin. 23.5 in (60 cm). 1962-1964. Scarcity: 4. $75-$125.

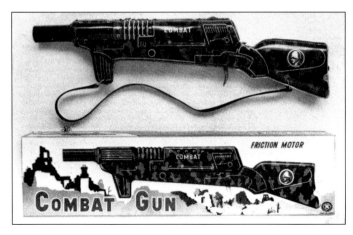

#3319 **Combat Gun** with sparking. Friction, tin. 19 in (48 cm). 1964-1967. Scarcity: 5. $50-$100.

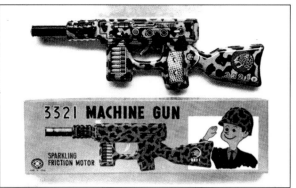

#3321 **Machine Gun** with sparking. Friction, tin. 14 in (36 cm). 1964-1968. Scarcity: 4. $50-$100.

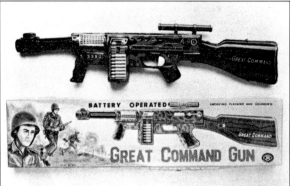

#3350 **Great Command Gun** with smoke, sound, and flashing lights. Battery operated, tin. 23.5 in (60 cm). 1964-1966. Scarcity: 6. $100-$175.

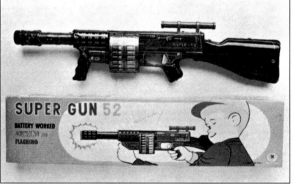

#3352 **Super Gun 52** with flashing light and sound. Battery operated, tin. 24 in (61 cm). 19651966. Scarcity: 6. $75-$150.

#3425 **Great Machine Gun** with ack-ack sound, flashing light, and rotating magazine belt. Battery operated, tin with plastic. 23.25 in (59 cm). 1966-1967. Scarcity: 6. $125-$200.

Jeeps   163

#3631 **Double Barreled Machine Gun 31** with twin barrels, turning flashing light, and machine gun sound. Battery operated, tin. 19.5 in (50 cm). 1968-1970. Scarcity: 5. $75-$150.

Cannons & Guns – Not Pictured

#1858 **Cannon**. Long barrel cannon with shells on six wheels, two of which are detachable. Tin. 13 in (33 cm). 1958. Scarcity: 4. $75-$150.

#1927 **Machine Gun** with revolving magazine. Friction, tin. 19.5 in (50 cm). 1959. Scarcity: 5. $50-$100.

#3047 **Eagle Machine Gun w/ Flash of Light**. Light flashes when gun is fired. Battery operated, tin. 24 in (61 cm). 1959-1964. Scarcity: 5. $50-$100.

#3064 **Jaguar Machine Gun with Flash Light**. Light flashes when gun with Jaguar litho is fired. Battery operated, tin. 21 in (53 cm). 1960(57)-1962. Scarcity: 5. $50-$100.

#3073 **Baby Machine Gun OK-20 with Sparkling**. OK-20 machine gun with sparking. Friction, tin. 10 in (25 cm). 1960-1962. Scarcity: 4. $40-$75.

#3672 **Turret Army Jeep**. Comic style jeep with driver and rotating machine gun and gunner. Battery operated, mystery action, tin with plastic. 12.25 in (31 cm). 1968-1980s. Also sold as Desert Patrol Jeep. Scarcity: 3. $100-$175.

Jeeps

#3381 **Army Jeep**. Comical jeep with rotating light in rear engine. Jeep stops, lifts up, turns as driver's face turns red. Battery operated, tin with vinyl head on driver. 15.25 in (39 cm). 1965-1969. Scarcity: 5. $100-$200.

#3464 **Brake Jeep** with braking action and driver's face turns red. Battery operated, tin with vinyl. 11 in (28 cm). 1966. Scarcity: 7. $100-$200.

#3742 **Army Jeep** with driver, radio, antenna, and machine gun with flashing light. Battery operated, mystery action, plastic with tin. 9.25 in (23 cm). 1969-1978. Scarcity: 3. $40-$75.

Tanks

#4293 **T.V. Combat Jeep** with driver, siren, and battlefield TV screen. Battery operated, mystery action, tin with plastic. 9.25 in (23 cm). 1974-1980s. Scarcity: 4. $50-$100.

Trade journal ad, February 1964.

#1600 **M1 Electric Tank** with crank elevated cannon and rubber treads. Battery operated, tin. 7.5 in (19 cm). 1956-1960. Scarcity: 4. $75-$125.

#4358 **Turret Navy Jeep**. Comic style jeep with driver, rotating machine gun, and gunner. Battery operated, mystery action, tin with plastic. 12.25 in (31 cm). 1975-1980s. Scarcity: 4. $75-$150.

#4548 **Electro Computer Vehicle** with six plastic disks (cams) to control programmable steering. Missiles can be spring fired from launcher. Battery operated, plastic. 7.5 in (19 cm). 1977-1980s. Scarcity: 4. $40-$75.

Jeeps – Not Pictured

#3062 **New Jeep** with fold down windshield and antenna. Friction, tin. 8.5 in (22 cm). 1960-1962. Scarcity: 3. $75-$125.

Tanks 165

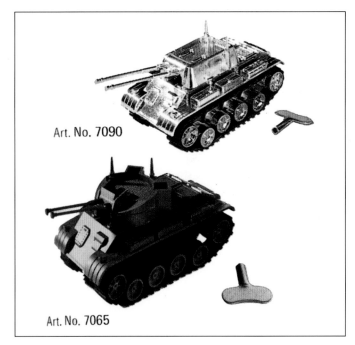

Baby Tank M-7. Friction, tin. 3 in (8 cm). 1950s. Scarcity: 3. *Courtesy of Barbara Moran.* $20-$30.

(top) #7090 **Mini Tank Silver "Archery"** silver plating with sparking action. Windup, plastic. 5.75 in (15 cm). 1972. Scarcity: 5. $30-$50.
(bottom) #7065 **Mini Tank "Hunter"** with sparking action. Windup, plastic. 5.75 in (15 cm). 1971-1972. Scarcity: 4. $30-$50.

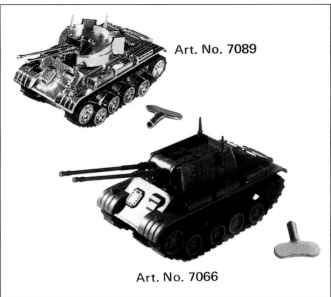

#4146 **M-12 Tank** with forward action, rotating turret, and flashing gun. Battery operated, plastic. 10.75 in (27 cm). 1973-1978. Scarcity: 3. $75-$125.

(top) #7089 **Mini Tank Silver "Hunter"** silver plating with sparking action. Windup, plastic. 5.75 in (15 cm). 1972. Scarcity: 5. $30-$50. (bottom) #7066 **Mini Tank "Archery"** with sparking action. Windup, plastic. 5.75 in (15 cm). 1971. Scarcity: 5. $30-$50.

#1519 **M 15 Tank** with sparkling and sound. Friction, tin. 7.5 in (19 cm). 1955-1960. Scarcity: 4. $60-$100.

#4305 **M-05 Caterpillar Tank** with long lever control and shooting missile. Battery operated, plastic with tin. 11 in (28 cm). 1975-1977. Scarcity: 5. $100-$150.

166 Military

#3216 **M-16 Tank**. 3216 tank with sparking and detonation sound. Friction, tin. 7.5 in (19 cm). 1962-1967. Scarcity: 3. $40-$75.

(top left) #3969 **M-39 Tank** with engine sound. Friction, plastic. 8.5 in (22 cm). 1972. Scarcity: 6. $75-$125. (top right) #3624 **M-24 Sparkling Tank** with sparking. Windup, tin. 9.25 in (23 cm). 1968-1972. Scarcity: 4. $75-$125. (bottom left) #3528 **M-28 Silver Tank** with sparking. Friction, tin. 7.25 in (18 cm). 1967-1969. Scarcity: 3. $40-$75. (bottom center) #1637 **M-25 Sparkling Tank** with sparking. Friction, tin. 4 in (10 cm). 1956-1964. Scarcity: 4. $25-$50. (bottom right) #3055 **M-55 Tank** with sparking. Friction, tin. 6.25 in (16 cm). 1960-1964. Scarcity: 4. $50-$75.

#4033 **M-33 Missile Tank** with six missiles, lever controlled direction, and missile launch. Battery operated, plastic and tin. 12.75 in (32 cm). 1972-1980s. Scarcity: 3. $75-$150.

(left) #4088 **Silver Tank M-88** with rubber treads, moving turret, and sparking gun. Windup, plastic and tin. 10.5 in (27 cm). 1972-1974. Scarcity: 4. $100-$150. (right) #4022 **M-22 Tank** with rubber treads, moving turret, and sparking gun. Windup, plastic and tin. 10.5 in (27 cm). 1972-1975. Scarcity: 5. $100-$150.

#3340 **Shooting Tank M-40** with 4-position remote for forward/reverse, rotating turret and firing rubber tipped darts. R/C battery operated, tin. 11.25 in (29 cm). 1964-1976. Scarcity: 3. $100-$150.

#3134 **M-34 Tank** with rubber treads and sparking. Windup, tin. 7.5 in (19 cm). 1962-1967. Scarcity: 3. $50-$100.

Tanks 167

Art. No. 4681
B/O R/C M-81 Tank

Art. No. 4640
B/O R/C M-40 Tank

Art. No. 4658
B/O R/C MS-58 MISSLE TANK

(left) **#4681 M-81 Tank with Flashing Gun** with dual levers, forward/reverse, rotating turret, and flashing gun. R/C battery operated, tin and plastic. 10.5 in (27 cm). 1977-1980s. Scarcity: 4. $75-$125. (center) **#4640 M-40 Tank** with forward/reverse, rotating turret, and shooting dart. R/C battery operated, tin and plastic. 10.5 in (27 cm). 1977-1980s. Scarcity: 4. $75-$125. (right) **#4658 MS-58 Missile Tank** with forward/reverse, rotating turret, and shooting missiles. R/C battery operated, plastic with tin. 11 in (28 cm). 1977-1980s. Scarcity: 4. $75-$125.

#3445 **Missile Tank M-45** with two missile rubber tipped plastic darts. Friction, tin. 7.5 in (19 cm). 1966. Scarcity: 7. $100-$150.

#3358 **M-58 Tank** with rubber treads, sparking, and opening hatch with soldier. Windup, tin with plastic. 5.5 in (14 cm). 1966-1970. Scarcity: 4. $40-$75.

(left) #3458 **Missile Tank MS-58** with 4-position control for direction, turret, and shooting rubber tipped missile darts. R/C battery operated, tin. 8.75 in (22 cm). 1966-1973. Scarcity: 4. $75-$125. (right) #3603 **M-103 Tank** with machine gunner, flashing gun, and rubber treads. Battery operated, tin with plastic. 7.25 in (18 cm). 1968-1972. Scarcity: 3. $50-$100.

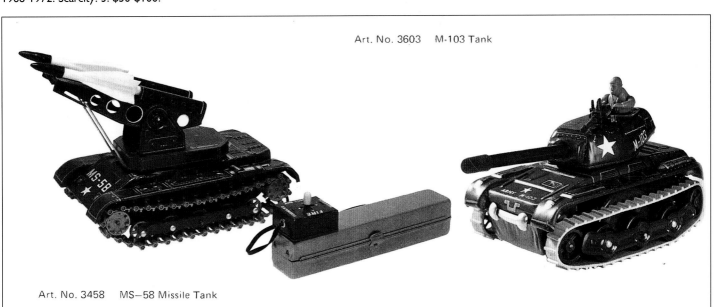

Art. No. 3603   M-103 Tank

Art. No. 3458   MS-58 Missile Tank

168   *Military*

#3280 **Silver Tank M-80** with sparking. Friction, tin. 7.25 in (18 cm). 1963-1966. Scarcity: 3. $40-$75.

#4258 **MS-58 R/C Missile Tank** with forward, reverse, turning, and firing remote control. R/C battery operated, tin with plastic. 8.25 in (21 cm). 1974-1976. Scarcity: 5. $75-$125.

#4181 **M-81 Tank** with forward/reverse action, rotating turret, and flashing gun. R/C battery operated, tin with plastic. 11.25 in (29 cm). 1973-1976. Scarcity: 4. $75-$125.

#3471 **M-71 Tank** with turning turret, shooting sound, and flashing gun. R/C battery operated, tin. 11.25 in (29 cm). 1966-1970. Scarcity: 5. $125-$175.

#3599 **Army Tank M-99** with antenna lever control and flashing light. Battery operated, plastic with tin. 18 in (46 cm). 1968-1980s. Scarcity: 2. $100-$150.

#4285 **MS-85 Missile Armored Vehicle** with driver in cockpit, long lever controls, and rotating nine missile launcher. Battery operated, tin with plastic. 13.5 in (34 cm). 1974-1976. Scarcity: 5. $150-$250.

(left) #4677 **M-77 Tank with Smoke** with antenna lever controlled direction and smoking gun barrel. Battery operated, plastic and tin. 9.75 in (25 cm). 1978-1980s. Scarcity: 5. $75-$125.
(right) #4688 **M-88 Tank with Flashing Gun** with antenna lever controlled direction and flashing gun barrel. Battery operated, plastic and tin. 9.75 in (25 cm). 1978-1980s. Scarcity: 5. $50-$100.

Tanks   169

#3390 **M-90 Shooting Tank** with rubber treads, revolving turret, and shooting darts. Windup, tin with plastic. 10.75 in (27 cm). 1965-1969. Scarcity: 5. $100-$150.

#3496 **M-96 Super Tank** with left and right turning, stopping, shooting sound, and flashing lights. Battery operated, tin. 17 in (43 cm). 1967. Scarcity: 7. $200-$300.

#3169 **M-120 Tank** with rubber treads and blinking cannon light. Battery operated, tin. 7.5 in (19 cm). 1962-1967. Scarcity: 4. $75-$125.

#4738 **Radicon Panther Tank** with single channel radio control. R/C battery operated, plastic. 12 in (30 cm). 1979-1980s. Scarcity: 5. $75-$125.

#3342 **Super Tank M-342** with authentic tread, flashing light, and rotating turret. Battery operated, tin. 16.5 in (42 cm). 1964-1967. Scarcity: 5. $150-$250.

#3512 **Radicon Tank** M-3512 with antenna and push button radio control. R/C battery operated, tin. 17 in (43 cm). 1967-1973. Scarcity: 5. $150-$250.

Tanks – Not Pictured

#1595 **M 12 Tank** with rubber treads and sparking. Windup, tin. 7.5 in (19 cm). 1955-1960. Scarcity: 4. $60-$100.

#1770 **Tank M-18**. Battery operated, mystery action, tin. 7.5 in (19 cm). 1957-1958. Scarcity: 5. $75-$125.

#1813 **X-1018 Tank** with antenna and lighted gun. Battery operated, non-fall, tin. 8.5 in (22 cm). 1957-1961. Scarcity: 4. $75-$125.

#1820 **Tank M-58** with rubber treads. Battery operated, tin. 7 in (18 cm). 1957-1961. Scarcity: 4. $75-$125.

#1911 **M-22 Tank** with sparking, antenna, and rubber treads. Windup, tin. 4 in (10 cm). 1958. Scarcity: 5. $50-$75.

#3157 **Missile Tank** with rubber treads and missile platform that fires finned plastic missile. Battery operated, tin with plastic. 7 in (18 cm). 1961-1962. Scarcity: 5. $100-$150.

#3651 **Caterpillar Tank M-1** with lever antenna control, flashing gun barrel, and sound. Battery operated, tin. 9.25 in (23 cm). 1968-1971. Scarcity: 4. $125-$175.

#4127 **MS-27 Missile Tank** with manual firing missiles and rubber treads. Pull toy, plastic. 8.5 in (22 cm). 1973-1974. Scarcity: 5. $40-$75.

#4271 **Radicon Missile Tank** with radio remote control of direction and firing of four missiles (comes with six missiles). R/C battery operated, plastic with tin. 14.5 in (37 cm). 1975-1977. Scarcity: 5. $125-$225.

#4803 **M-80 Tank** with rotating turret and flashing lights. Battery operated, tin and plastic. 9.5 in (24 cm). 1980. Scarcity: 5. $50-$75.

#7015 **GO GO Tank** with sparking. Windup, plastic with tin. 5.5 in (14 cm). 1970-1972. Scarcity: 3. $25-$45.

Trucks

#3283 **Japanese Defense Army Truck**. Chevrolet truck with vinyl camouflage top. Friction, tin. 11.5 in (29 cm). 1964-1965. Scarcity: 8. $500-$900.

#1802 **Radarscope Car** with radar scope and rotating antenna. Battery operated, tin. 11.5 in (29 cm). 1957-1961. Scarcity: 6. $175-$350.

#3288 **Army Truck**. Chevrolet army truck with vinyl camouflage top. Friction, tin. 11.5 in (29 cm). 1964-1967. Scarcity: 7. $400-$800.

#4136 **MS-36 Missile Truck** with missile launcher that fires missiles. Pull toy, tin and plastic. 10.75 in (27 cm). 1973-1974. Scarcity: 5. $40-$80.

Trucks – Not Pictured

#3045 **Pressure Control Half Truck**. Half-track style vehicle with full track. Controlled by pressing antenna rod. Battery operated, tin. 8 in (20 cm). 1959-1962. Scarcity: 5. $75-$150.

Other

#1817 **Radar Station** with radar scope, rotating radar antenna, lights, and warning sounds. Battery operated, tin. 10 in (25 cm). 1957-1961. Scarcity: 7. $150-$300.

#1832 **A.A. Searchlight Car** with pull out antenna, rotating searchlight, and tin military figure. Battery operated, tin. 11 in (28 cm). 1958-1960. Scarcity: 6. $100-$200.

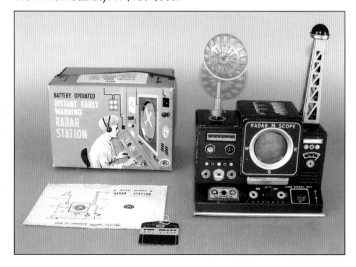

Motorcycles

Kiddy Motorcycle with Sidecar with celluloid rider and sidecar. Windup, tin and celluloid. 5.5 in (14 cm). 1930s. Scarcity: 7. $250-$500.

Military Motorcycle with military rider on camouflage cycle A-754. Windup, tin. 8 in (20 cm). 1930s. Scarcity: 8. $750-$1,500.

Round Motor-Cycle Cable Rider. Police motorcycle rider follows spring-like cable around irregular path. Windup, tin. 5 in (13 cm). 1950s. Scarcity: 4. $275-$400.

Motorcycle with child-like riders in military dress with doll and teddy bear. Windup, tin. 1930s. Scarcity: 10. *Courtesy of Morphy Auctions.* $6,000-$9,000.

#1731 **Expert Motor-Cyclist** with tin rider that rides, stops, dismounts, and remounts cycle. Battery operated, tin. 11.5 in (29 cm). 1957-1958. Scarcity: 6. $500-$900.

172   Motorcycles

#1772 **Police Motor Cycle** with helmeted police rider in green uniform. Friction, tin. 8.5 in (22 cm). 1957-1958. Scarcity: 5. $400-$800.

#1772B **General Motor Cycle** with rider and detonation sound. Friction, tin. 8.5 in (22 cm). 1958. Scarcity: 6. $500-$800.

#1859 **High Way Patrol** with tin rider that rides, stops, dismounts, and remounts cycle. Battery operated, tin. 11.5 in (29 cm). 1958-1962. Reintroduced as #3473 in 1966. Scarcity: 5. $450-$850.

#3019 **Fire Patrol Motor Cycle** with tin Fire Chief rider that rides, stops, dismounts, and remounts cycle. Battery operated, tin. 11.5 in (29 cm). 1960-1964. Reintroduced as #3474 in 1966. Scarcity: 8. $600-$1,000.

#3113 **World Champion Motorcycle** with tin racing rider that rides, stops, dismounts, and remounts cycle. Battery operated, tin. 11.5 in (29 cm). 1960-1964. Scarcity: 8. $700-$1,200.

Motorcycles 173

#3199 **Police Motorcycle**. Green Highway Patrol motorcycle with tin police rider that rides, stops, dismounts, and remounts cycle. Battery operated, tin. 11.75 in (30 cm). 1962-1964. Scarcity: 7. $700-$1,200.

#3410 **Grand Prix Racer**. Racing cycle number 10 with vinyl head driver. Friction, tin with vinyl. 11.5 in (29 cm). 1966-1967. Scarcity: 6. $125-$225.

(left) #3807 **Fire Chief on Motor Cycle** with Fireman cyclist that goes and stops with siren and flashing headlight. Battery operated, tin and plastic. 11.5 in (29 cm). 1970-1973. Scarcity: 6. $200-$350.
(right) #3667 **Siren Patrol Motorcycle** with vinyl Police cyclist that goes and stops with large siren and flashing headlight. Battery operated, tin with plastic. 11.5 in (29 cm). 1968-1973. Scarcity: 3. $200-$300.

#3341 **Motor Cycle Racer**. Racing cycle number 27 with vinyl head driver. Cycle turns quickly with sound. Battery operated, tin with vinyl. 11.5 in (29 cm). 1964-1967. Scarcity: 5. $125-$225.

#3473 **High Way Patrol Cycle** with tin Police rider that rides, stops, dismounts, and remounts cycle. Battery operated, tin. 11.5 in (29 cm). 1966. Scarcity: 5. $400-$800.

Motorcycles

#4075B **Police Motorcycle** with Police rider, sound, headlight, steering, and forward/reverse. R/C battery operated, plastic. 9 in (23 cm). 1972. Scarcity: 6. $100-$200.

#3899 **Police Patrol Motorcycle** with vinyl rider and flashing lights. R/C battery operated, tin with vinyl. 11.5 in (29 cm). 1971-1973. Scarcity: 5. $175-$350.

Motorcycles – Not Pictured

Motor Cycle. Tin. 4.5 in (11 cm). 1940s. Scarcity: 7. $300-$500.

MP Motorcycle with rider and siren sound. Friction, tin. 7 in (18 cm). 1950s. Scarcity: 6. $200-$400.

Police Motorcycle with helmeted police rider and machine gun on Indian cycle. Friction, tin. 5 in (13 cm). 1950s. Scarcity: 6. $250-$500.

Side Car with police on motorcycle and in sidecar. Friction, tin. 1950s. Scarcity: 7. $400-$800.

#3113B **Champion Motorcycle**. Yellow "Champion" No.27 with tin racing rider that rides, stops, dismounts, and remounts cycle. Battery operated, tin. 11.5 in (29 cm). 1961. Scarcity: 8. $700-$1,200.

#3474 **Fire Patrol Cycle** with tin Fire Chief rider that rides, stops, dismounts, and remounts cycle. Battery operated, tin. 11.5 in (29 cm). 1966. See similar #3019. Scarcity: 8. $500-$1,000.

#4048 **Police Motor Cycle** with vinyl Police driver and engine noise. Friction, plastic. 9 in (23 cm). 1973. Scarcity: 5. $75-$150.

#4375 **Police Side Car** with Police driver. Battery operated, plastic with tin. 6.5 in (17 cm). 1975. Scarcity: 6. $75-$175.

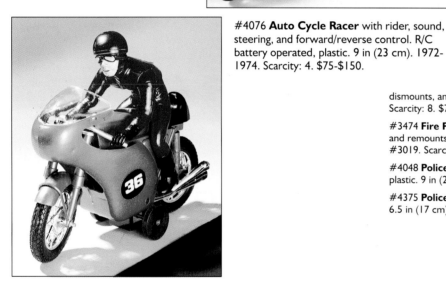

#4076 **Auto Cycle Racer** with rider, sound, steering, and forward/reverse control. R/C battery operated, plastic. 9 in (23 cm). 1972-1974. Scarcity: 4. $75-$150.

#4047 **Speed Racer Motor Cycle** with vinyl driver and engine noise. Friction, plastic. 9 in (23 cm). 1973. Scarcity: 5. $75-$150.

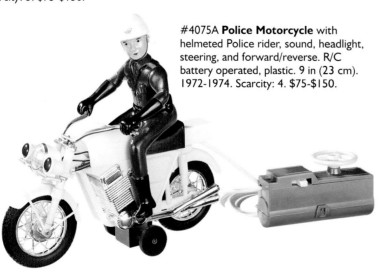

#4075A **Police Motorcycle** with helmeted Police rider, sound, headlight, steering, and forward/reverse. R/C battery operated, plastic. 9 in (23 cm). 1972-1974. Scarcity: 4. $75-$150.

#4384 **Police Motor Cycle** with Police driver. Windup, plastic with tin. 6.5 in (17 cm). 1975-1980. Scarcity: 5. $50-$100.

People

Girl and Baby. Girl lifts jointed baby in the air. Windup, celluloid. 4.5 in (11 cm). 1950s. Scarcity: 6. $100-$200.

Kojinbutsu (Mr. Nice Guy). Political satire of Teddy Roosevelt being bothered by a bee (Japan). Windup, celluloid. 9 in (23 cm). 1930s. Also produced in 7 inch version Scarcity: 8. $800-$1,400.

Skating Couple. Couple ice skating. Windup, celluloid. 5 in (13 cm). 1948. Scarcity: 6. $150-$250.

Serpent Charmer. Battery operated, tin. 7 in (18 cm). 1960s. Scarcity: 8. $500-$800.

Xylophone Player plays musical xylophone. Windup, celluloid and tin. 5.75 in (15 cm). 1952. Scarcity: 5. $150-$300.

#1064 **Tumbling Doll** with long rotating arms that cause doll to tumble. Windup, celluloid. 6 in (15 cm). 1940s. Scarcity: 3. $75-$125.

176　People

#1073A **Baby Porter**. Black boy porter with luggage. Windup, celluloid. 3.5 in (9 cm). 1940s. Scarcity: 6. *Bill & Stevie Weart Collection*. $200-$400.

#1500 **Magic Snow Man** with lighted eyes, waving arm, and ball blowing top hat. Battery operated, cloth over tin. 11.25 in (29 cm). 1955. Scarcity: 3. $150-$225.

#1073B **Porter**. Black porter carrying luggage. Windup, celluloid. 5 in (13 cm). 1940s. Scarcity: 7. $350-$700.

#3487 **Open Sleigh**. Eskimo driving sled pulled by set of two moving dogs with bell sound. Battery operated, mystery action, tin with plastic. 15.5 in (39 cm). 1967-1969. Scarcity: 7. $300-$500.

#3487s **Open Sleigh (mock-up)**. This sample has Eskimo driving sled pulled by set of three moving dogs. Production model had two dogs. 1967. Scarcity: 10. *Courtesy of Japan Toys Museum*.

#1099 **Skipping Couple**. Couple skipping rope. Windup, celluloid. 5 in (13 cm). 1940s. Scarcity: 6. $175-$325.

#3911 **Baby Cyclist with Dog**. Baby on tricycle with running dog and ringing bell. Windup, plastic with tin. 5.75 in (15 cm). 1971-1973. Scarcity: 4. $30-$50.

#4264 **Kiddy Tricycle** with kiddy driver on tricycle with ringing bell. Battery operated, plastic with tin. 8 in (20 cm). 1974-1976. Scarcity: 4. $50-$100.

#660/1090 **Ice Cream Carrier**. Celluloid boy pedaling on tin ice cream cart with painted stenciled lettering. Windup, celluloid and tin. 4 in (10 cm). 1940s. Scarcity: 5. $150-$250.

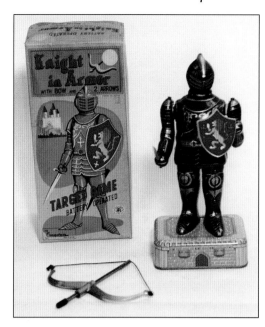

#1795 **Knight in Armour Target Game** with crossbow and two arrows. Battery operated, tin. 12.25 in (31 cm). 1957. Scarcity: 7. $225-$450.

Skating Santa. Windup. 1940s. Scarcity: 7. *Courtesy of Barbara Moran.* $400-$800.

(top) #1776 **B-Z Ice Cream Vender** with tin driver on ice cream cart. Battery operated, mystery action, tin. 8 in (20 cm). 1957-1958. Scarcity: 10. $1,250-$2,500. (bottom) #5772 **Ice Cream Vendor mini replica**. Windup, tin. 1997-1998. Scarcity: 2. $15-$25.

(top) #1808 **Walking Knight in Armour** walks with shield and sword. R/C battery operated, tin. 10 in (25 cm). 1957-1961. Scarcity: 10. $1,000-$2,000. (bottom) #5780 **Walking Knight in Armour mini replica**. Windup, tin. 1997-1998. Scarcity: 2. $10-$20.

178 People

#4090 **Santa Claus on Hand Car** moves in rowing manner on hand car with ringing bell. Battery operated, mystery action, tin and plastic. 10 in (25 cm). 1973-1974. Scarcity: 4. $100-$200.

#3488 **Santa Claus on Reindeer Sleigh**. Santa Claus with single reindeer, bell, and blinking lantern. Battery operated, mystery action, tin with plastic. 18 in (46 cm). 1967-1970. Scarcity: 6. $200-$400.

#3991 **Santa Claus on Scooter** with flashing head and tail lights. Battery operated, mystery action, tin and plastic. 9.5 in (24 cm). 1972-1973. Scarcity: 4. $75-$175.

Soldier kneeling with rifle. Windup, celluloid. 6.25 in (16 cm). 1940s. Scarcity: 7. $200-$400.

#4241 **Santa Copter** space ship with helicopter blades, flashing light, sound and Santa pilot. Battery operated, non-fall, tin with plastic. 9.5 in (24 cm). 1974-1976. Scarcity: 2. $75-$150.

People   179

Sparkling Machine with Gunner kneeling and firing large machine gun on wood base. Windup, celluloid and tin. 10 in (25 cm). 1930s. Scarcity: 8. *Courtesy of Keith Spurgeon.* $600-$1,000.

Shingun No.2 soldier with rifle. Windup, celluloid. 8.75 in (22 cm). 1930s. Scarcity: 6. *Courtesy of Rex & Kathy Barrett.* $250-$500.

Rooting Tooting Cowboy. Cowboy walks forward with moving head and arms with guns. Windup, tin and celluloid. 8.25 in (21 cm). 1949. Similar all celluloid toy sold as "Cowboy With Two Guns." Scarcity: 5. $100-$200.

#3302 **Overland Stage Coach** with tin wagon driver, galloping sound, and moving horses. Battery operated, tin with plastic. 17.75 in (45 cm). 1964-1969. Scarcity: 5. $125-$250.

180   People

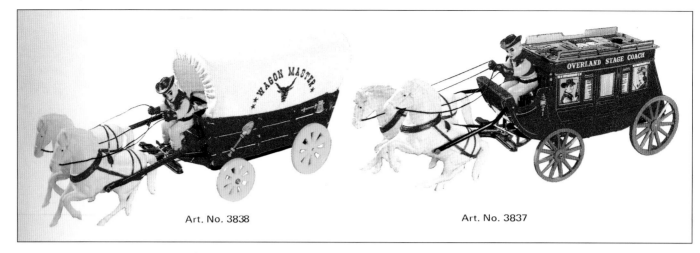

Art. No. 3838   Art. No. 3837

(left) **#3838 Wagon Master** with plastic wagon driver, galloping sound, and moving horses. Battery operated, plastic with tin. 18.25 in (46 cm). 1970-1973. Scarcity: 4. $75-$125. (right) **#3837 Overland Stage Coach** with plastic wagon driver, galloping sound, and moving horses. Battery operated, tin with plastic. 17.75 in (45 cm). 1970-1973. Reissue of item #3302. Scarcity: 4. $75-$125.

#3303 Western Ranger. Vinyl cowboy on tin horse with galloping sound. R/C battery operated, tin with vinyl. 10.5 in (27 cm). 1963-1967. Scarcity: 7. $200-$400.

#3345 **Covered Stage Wagon/Wagon Master** with driver, galloping sound, and two moving horses. Battery operated, tin, plastic and vinyl. 18.25 in (46 cm). 1964-1969. Scarcity: 6. $150-$300.

Horse and Cart. Cowboy standing on cart pulled by jumping horse. Windup, celluloid and tin. 10 in (25 cm). 1950s. Scarcity: 4. $175-$350.

Indian on jumping horse. Windup, celluloid. 7 in (18 cm). 1950s. Scarcity: 4. $125-$250.

Roll Around (Boy or Girl). Celluloid boy or girl with bell rolling in tin hoop. Windup, tin and celluloid. 6.5 in (17 cm). 1950s. Scarcity: 6. $200-$400.

Three Dancing Girls with three girls dancing on platform. Windup, celluloid and tin. 5 in (13 cm). 1950s. Scarcity: 8. $300-$600.

Walk Santa Claus. Walking Santa Claus with lead legs. Windup, celluloid. 7 in (18 cm). 1940s. Scarcity: 8. $400-$800.

#1031 **Dancing Couple** dances and turns realistically. Windup, celluloid and tin. 4.5 in (11 cm). 1948. Scarcity: 3. $75-$150.

#1069 **Cow-Boy.** Cowboy on galloping horse with large front wheel. Windup, celluloid and tin. 6 in (15 cm). 1940s. Scarcity: 6. $200-$400.

#1091 **Donkey With Rider.** Boy sitting on walking donkey with moving head. Windup, celluloid. 5.5 in (14 cm). 1940s. Scarcity: 4. $100-$200.

#1292C **Bucking Bronco.** Celluloid cowboy sits on horse that rears back on hind legs and tail. Windup, celluloid and tin. 8 in (20 cm). 1952-1960. Scarcity: 2. $50-$100.

People – Not Pictured

Cow Boy with Two Guns. Cowboy raises both arms with gun in each hand. Windup, celluloid. 5 in (13 cm). 1940s. Scarcity: 4. $75-$150.

Cowboy on jumping horse. Windup, celluloid. 7 in (18 cm). 1950s. Scarcity: 4. $125-$250.

Dandy Girl. Standing girl applying makeup with mirror. Windup, celluloid. 5 in (13 cm). 1950s. Scarcity: 5. $250-$450.

#1793 **Ice Cream Vender.** 3-wheel ice cream cart with polar bear graphics and driver. Windup, tin. 8 in (20 cm). 1957. Scarcity: 7. $200-$400.

#1834 **Folk Dance** with boy and girl dancers magnetically moving around on tin base. Battery operated, tin. 6 in (15 cm). 1958-1960. Scarcity: 6. $100-$150.

#3108 **My Baby Santa Claus.** Baby in Santa outfit with package and ringing bell. Windup, cloth, vinyl and tin. 4.5 in (11 cm). 1960-1962. Scarcity: 2. $20-$40.

Space

1971 catalog page.

1970 catalog cover.

Ray Guns

Ray Gun with sparking. Friction, tin. 8.5 in (22 cm). 1950s. Scarcity: 5. $100-$150.

Art. No. 2061

Art. No. 2062

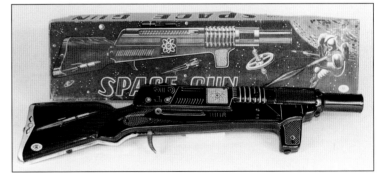

#1916 **Space Gun** with sparking. Friction, tin. 19 in (48 cm). 1958-1962. Scarcity: 5. $100-$200.

(top) #2061 **Space Ray Gun** with sparking. Friction, tin with plastic. 10 in (25 cm). 1960s-1980s. Scarcity: 4. $75-$125.
(bottom) #2062 **Space Ray Gun** with sparking. Friction, tin with plastic. 12 in (30 cm). 1960s-1980s. Scarcity: 4. $75-$125.

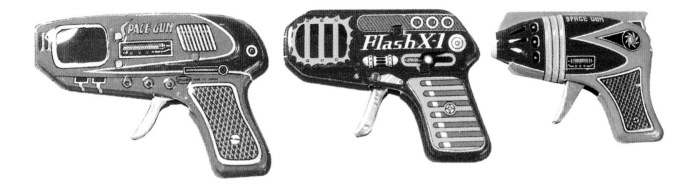

Art. No. 2066　　　Art. No. 2065　　　Art. No. 2064

(left to right) #2066 **Space Gun** with sparking. Friction, tin. 6 in (15 cm). 1960s-1980s. Scarcity: 4. $35-$65.
#2065 **Flash Gun** with sparking. Friction, tin. 5 in (13 cm). 1960s-1980s. Scarcity: 4. $35-$65. #2064 **Mini Space Gun** with sparking. Friction, tin. 4 in (10 cm). 1960s-1980s. Scarcity: 4. $35-$65.

Art. No. 2067

Art. No. 2068

(left) #2068 **Astroray Gun** with sparking. Friction, tin. 9.5 in (24 cm). 1960s-1980s. Scarcity: 4. $50-$100.
(right) #2067 **Laser Gun** with sparking. Friction, tin. 8 in (20 cm). 1960s-1980s. Scarcity: 4. $45-$90.

#3102 **Silver Eagle Machine Gun**. Space machine gun with flashing barrel and sound. Space graphics on box. Battery operated, tin. 24 in (61 cm). 1960-1964. Scarcity: 6. $75-$150.

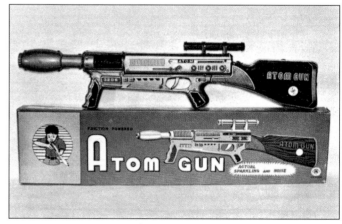

#3206 **Atom Gun**. Atom Gun 3206 with scope and sparking. Friction, tin. 24 in (61 cm). 1962-1966. Scarcity: 5. $75-$150.

#2069b **Cosmic Rays Gun** with sparking. Friction, tin. 13.5 in (34 cm). 1960s-1980s. Scarcity: 4. $75-$125.

#3763 **Sparkling Space Gun S-63** with see through sparking. Friction, plastic. 9 in (23 cm). 1969-1973. Scarcity: 3. $20-$40.

#4749 **Space Gun with Sound** and flashing light. Battery operated, plastic. 11 in (28 cm). 1979. Scarcity: 4. $30-$50.

Ray Guns – Not Pictured

#1831 **Atomic Ray Gun** with sound. Battery operated, tin. 18 in (46 cm). 1958-1960. Scarcity: 6. $125-$250.

Robots

Masudaya did not produce as many robots as some other Japanese toy companies. However, they are very famous for producing a series of five skirted robots over a period of seven years. Robot collectors have affectionately named this very collectible group of robots the "Gang of Five." Among the "Gang of Five," the first robot produced was the *Radicon Robot* in 1957. This was the second Radicon toy and a technically difficult robot to stamp and manufacture. The design was an extremely large robot for the time. The *Non-Stop Robot* followed the *Radicon Robot* in 1959. The *Non-Stop Robot* is referred to as the "Lavender Robot" because of its lavender color. It used the same body but had the function of non-stop action.

A special order from one of the American customers then led to production of a red color robot variation based on the non-stop lavender robot. Its name was *Giant Machine Man*. The quantity produced was very small, resulting in a very rare toy today.

In 1962, the *Giant Sonic Robot* was released. This robot also had non-stop action with a supersonic sound. In 1964, *Shooting Giant Robot* or Target Robot was added to the line. It came with a dart-shooting pistol. Hitting the target caused the robot to turn and go away from you before returning again.

#1715A **Space Commando** with light and lithographed remote control. R/C battery operated, tin. 7.5 in (19 cm). 1957-1960. Also produced in red color (+25%). Scarcity: 9. $1,500-$2,500.

Gang of Five. Complete set of original "Gang of Five" skirted robots. 15 in (38 cm). 1950-60s.

#1655 **Robot R-35**. R-35 walking robot with lighted eyes, moving arms, and lithographed remote control. R/C battery operated, tin. Silver or steel blue color. 7.75 in (20 cm). 1955/56-1960. Also sold under Linemar name. Scarcity: 5. *Courtesy of Smith House Toys.* $300-$600.

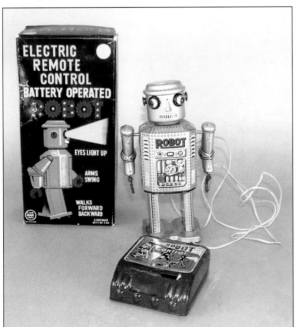

#5766 **Antique Robot R-35 mini replica**. Reissue of #5126. Windup, plastic and tin. 4.5 in (11 cm). 1997-1998. Scarcity: 1. $10-$20.

184 Space

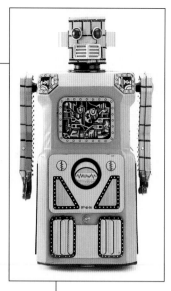

#3058a **Non Stop Robot (Lavender)**. Lavender colored robot was the second robot produced in what was to become the "Gang of Five." Machine Man was a variation of this robot. Battery operated, mystery action, tin. 15 in (38 cm). 1959. Scarcity: 8. $3,000-$6,000.

#5775 **Lavender Robot mini replica**. Windup, tin. 1997-1998. Scarcity: 1. $15-$25.

#5770 **Machine Man Robot mini replica**. Windup, tin. 1997-1998. Scarcity: 1. $15-$25.

Radicon Robot with antenna and radio control. The first of the Gang of Five robots. R/C battery operated, tin. 15 in (38 cm). 1957. Scarcity: 9. $6,000-$12,000.

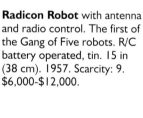

#3058b **Giant Machine Man**. OEM variation of Lavender robot that was the third robot produced in what was to become the "Gang of Five." Considered one of the rarest of all robots. Battery operated, mystery action, tin. 15 in (38 cm). 1959. Scarcity: 10. $25,000-$42,000.

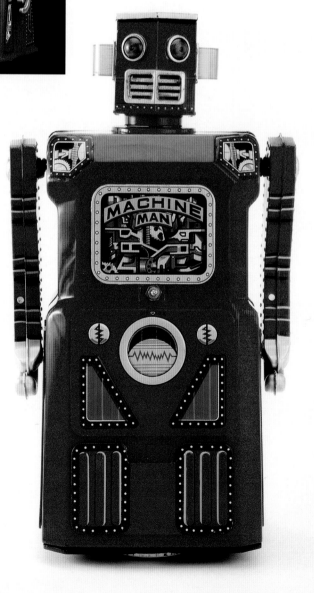

#5777 **Radicon Robot mini replica**. Windup, tin. 1997-1998. Scarcity: 1. $15-$25.

Robots 185

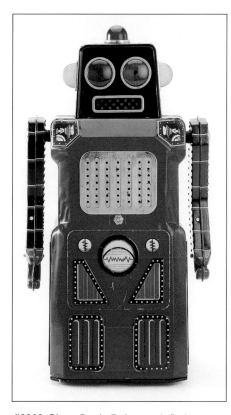

#3223 **Giant Sonic Robot** with flashing eyes and sound. Robot number four in the "Gang of Five" series. Battery operated, mystery action, tin. 15 in (38 cm). 1962-1964. Scarcity: 9. $5,000-$9,000.

#3357 **Shooting Giant Robot** or **Robot - Shoot Him (Target Robot)** with dart gun. Shoot the robot's target and it retreats from you. Robot number five in the "Gang of Five" series. Battery operated, tin. 15 in (38 cm). 1964-1967. Scarcity: 9. $4,500-$8,500.

#5778 **Sonic Robot mini replica**. Windup, tin. 1997-1998. Scarcity: 1. $15-$25.

(left) #5779 **X-9 Space Robot Car mini replica** with balls popping in plastic dome. Windup, tin and plastic. 4 in (10 cm). 1997-1998. Scarcity: 2. $10-$20. (right) #3121 **Robot Car X-9** with robot driver and ball blowing dome. Battery operated, non-stop, tin with plastic. 7.75 in (20 cm). 1961-1962. Scarcity: 7. $1,500-$2,500.

#5776 **Target Robot mini replica**. Windup, tin. 1997-1998. Scarcity: 1. $15-$25.

#3368 **Mighty 8 Robot** with swinging arms, sound, and swirling magic color wheel on chest. Battery operated, tin. 12 in (30 cm). 1964-1966. Scarcity: 8. $2,000-$4,000. Photo illustrates production robot on left and hand painted sample on right.

#4140 **R/C Missile Robot MR-45** with rotating body and missile firing. R/C battery operated, plastic. 14 in (36 cm). 1973-1975. Scarcity: 6. $350-$700.

186   Space

(left to right) #5125 **Forbidden Planet Robby Robot**. Miniature of 1956 movie robot. Windup, plastic. 4.5 in (11 cm). 1984. Scarcity: 2. $15-$25. #5217 **Robot YM-3**. Miniature of 1966 movie robot. Windup, plastic. 5 in (13 cm). 1985. Scarcity: 2. $15-$25. #5126 **Antique Robot R-35**. Miniature of original 1956 robot. Windup, plastic and tin. 4.5 in (11 cm). 1984. Scarcity: 2. $15-$25.

#5768 **Robby The Robot mini replica**. Reissue of #5125. Windup, plastic. 4.5 in (11 cm). 1997-1998. Scarcity: 1. $10-$20.

#8806 **Giant Robby Robot** from movie. Model required minor assembly and painting. Talks while light blinks in chest. Battery operated, vinyl. 24 in (61 cm). 1997-1998. Scarcity: 7. $200-$400.

#8542 **Metropolis UFA Maria Display Figure** from movie with lights in base. Battery operated, vinyl. 17.5 in (44 cm). 1985. Scarcity: 4. $100-$150.

#8448 **Robby - Forbidden Planet** from movie. Model required minor assembly and painting. Talks while light blinks in mouth. Battery operated, vinyl. 16 in (41 cm). 1984. Scarcity: 3. $125-$175.

Space Exploration and Travel   187

While Masudaya is famous for the skirted robots that came to be known as the "Gang of Five," these toys were not successful at the time—making them rare today. The real success in depicting the race for space using toys came with Masudaya's space ships and space explorers. Produced in a variety of shapes, these non-fall and mystery action toys were produced for many years. In fact, variations of the round saucer shaped toys were produced into the 1990s. Included here are a variety of rockets, space ships, explorers, and satellites.

#8501 **Robot YM-3** from movie. Model required minor assembly and painting. Talks while light blinks in chest. Battery operated, vinyl. 16 in (41 cm). 1985. Scarcity: 4. $150-$200.

Space Exploration and Travel

Trade journal ad, February 1953.

Trade journal ads, June 1958 and February 1963.

#1848 **Space Trip** with two space ship like racers that go in orbit. Battery operated, tin. 16.25 in (41 cm). 1958-1960. Scarcity: 7. $400-$700.

#1860 **Universe Traveler** Space station with radar antenna that sends space rocket car out and back with an over and under action. Windup, tin. 18 in (46 cm). 1958. Scarcity: 8. *Courtesy of Stan Luksenburg.* $400-$650.

Rocket Car X with lithographed figures in windows. Friction, tin. 6 in (15 cm). 1950s. Scarcity: 8. *Courtesy Smith House Toys.* $250-$400.

(left) #3803 **Space Jet X-3** with vinyl space pilot and jet noise. Friction, tin. 12.75 in (32 cm). 1970-1971. Scarcity: 6. $100-$175.
(right) #3713 **Space Patrol 713 w/ Boy** with vinyl head boy pilot and crank siren. Friction, tin with vinyl. 12.75 in (32 cm). 1969-1971. Scarcity: 7. $150-$275.

#3325 **Space Sightseeing Bus** with lithographed people in windows, sound, and lights. Battery operated, mystery action, tin. 13 in (33 cm). 1964-1967. Scarcity: 9. $700-$1,200.

#3852 **Space Car** SX-10 with vinyl space pilot. Battery operated, non-stop and non fall, tin with vinyl. 9.25 in (23 cm). 1970-1971. Scarcity: 6. $200-$400.

Space Exploration and Travel 189

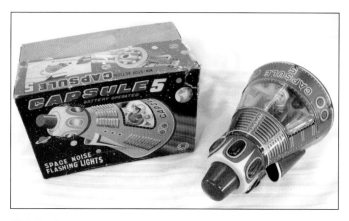

#3689 **Space Boy The Cyclist** with ringing bell. Windup, plastic with tin bell. 5.5 in (14 cm). 1968-1970. Scarcity: 3. $25-$45.

#3245b **Capsule 5** with tin astronaut pilot, sound, and flashing lights. Battery operated, mystery action, tin. 10.5 in (27 cm). 1962. Scarcity: 8. *Courtesy of Joe Morabito.* $200-$450.

#3764 **Space Boy Baby Car** with space boy driver and tin bell. Windup, plastic with tin bell. 5.5 in (14 cm). 1969. Scarcity: 7. $50-$100.

#3417 **Capsule No.7** with flashing lights and tin floating astronaut. Battery operated, mystery action, tin. 10 in (25 cm). 1966-1969. Scarcity: 5. $350-$550.

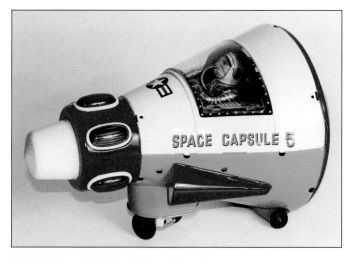

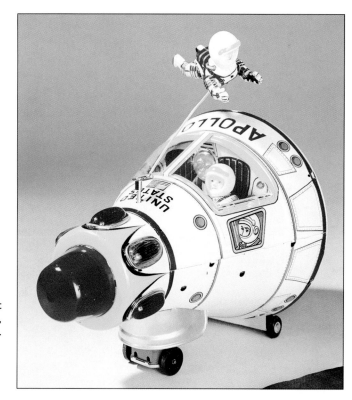

#3245a **Capsule 5** United States Space Capsule with tin astronaut pilot and flashing lights. Battery operated, mystery action, tin. 10.5 in (27 cm). 1962. Scarcity: 8. $250-$500.

#3784 **Apollo Spacecraft** with floating astronaut and flashing lights. Battery operated, mystery action, tin. 10 in (25 cm). 1970-1973. Scarcity: 4. $125-$250.

#3785 **Mini Apollo** with non fall action and vinyl pilot. Windup, tin with vinyl. 5.5 in (14 cm). 1970-1971. Scarcity: 3. $50-$100.

#3051 **Rocket Launching Base** with radar screen, rotating antenna, and rocket launcher. Battery operated, tin. 10 in (25 cm). 1959-1962. Scarcity: 7. $250-$500.

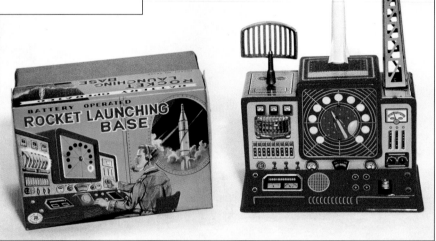

#3831 **NASA Control Center** displaying Lunar Module and Space Capsule with sound and flashing light. Battery operated, tin and plastic. 7.5 in (19 cm). 1970-1973. Scarcity: 4. $75-$150.

#1946 **Planet Explorer** with tin pilot and rotating antenna. Battery operated, non-fall, tin with plastic. 9 in (23 cm). 1959-1961. Scarcity: 4. *Courtesy of Joe Morabito*. $125-$250.

#3521 **Space Vehicle 3 Assortment**. Three different space vehicles from same body with pilots in clear cockpit. Friction, tin. 7 in (18 cm). 1967-1972. Scarcity: 4. $125-$250.

Space Exploration and Travel   191

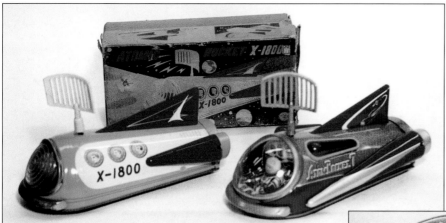

(left) #1883 **Atomic Rocket X-1800** with rotating plastic radar antenna and lithographed people in windows. Battery operated, non-fall, tin. 9 in (23 cm). 1958-1960. Scarcity: 7. $200-$400.
(right) #3178A **Atom Rocket 7** with pilot, flashing lights, and rotating plastic antenna. Battery operated, non-fall, tin. 9 in (23 cm). 1962-1969. Scarcity: 5. $200-$400.

#3178B **Atom Rocket 7** with pilot, flashing lights, and rotating plastic antenna. Color variation of 3178A. Battery operated, non-fall, tin. 9 in (23 cm). 1962-1969. Scarcity: 5. *Fineas J. Whoopie Collection*. $200-$400.

#3097 **Explorer** with large astronaut in cockpit and detonation sound. Friction, tin. 7.5 in (19 cm). 1960-1967. Scarcity: 7. *Courtesy of Cybertoyz*. $175-$350.

#4709 **Radicon Space Explorer**. One channel ray gun type controller with wheeled space vehicle. R/C battery operated, plastic. 8 in (20 cm). 1979-1980s. Scarcity: 5. $60-$125.

(left) #3414 **Gemini Rocket**. Apollo X-5 with floating tin astronaut. Battery operated, non-fall, tin. 9.5 in (24 cm). 1966-1969. Scarcity: 5. $200-$400. (right) #3304 **Moon Rocket** with astronaut on antenna platform. Battery operated, non-fall, tin with plastic. 9.25 in (23 cm). 1964-1972. Scarcity: 4. $150-$250.

192 Space

#3783 **Apollo Rocket** with floating astronaut and flashing lights. Battery operated, non-fall, tin with plastic. 9.5 in (24 cm). 1970-1979. Scarcity: 3. $75-$150.

#3247b **Friendship No. 7** with tin pilot, sound, and rotating (floating) tin astronaut. Battery operated, mystery action, tin. 10 in (25 cm). 1962-1964. Scarcity: 8. $300-$600.

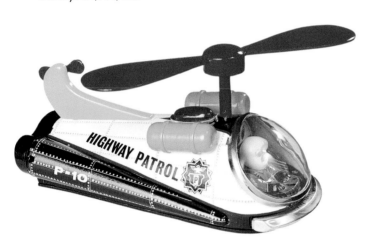

#4110 **Highway Patrol Helicopter**. Space ship with helicopter blades, flashing light, sound, and pilot. Battery operated, non-fall, tin with plastic. 9.5 in (24 cm). 1973-1979. Scarcity: 4. $75-$150.

#2963 **Apollo Target Game** with dart shooting gun and two darts. Plastic. 16.5 in (42 cm). 1970-1973. Scarcity: 4. $40-$100.

#4721 **Space Ranger No. 1** with rotating antenna, flashing lights, and blacked out cockpit. Battery operated, tin with plastic. 9.5 in (24 cm). 1979. Scarcity: 4. $75-$150.

#3035 **Moon Explorer** with astronaut in cockpit. Battery operated, mystery action, tin. 13.5 in (34 cm). 1959-1964. Scarcity: 6. *Courtesy of Cybertoyz.* $250-$450.

Space Exploration and Travel   193

#3043a **Sonicon Rocket** with tin pilot, rotating antenna, lights, and whistle sound direction control. Lt blue/dark blue/red. Battery operated, tin with plastic. 14 in (36 cm). 1959-1967. Scarcity: 7. *Courtesy of Cybertoyz.* $350-$600.

(left) #3679 **Space Patrol "Fire Bird"** with pilot and blinking light. Battery operated, mystery action, tin. 13.5 in (34 cm). 1968-1970. Scarcity: 6. $300-$600. (right) #3137 **Moon Ship Supersonic** with pilot. Battery operated, non-stop, tin. 13.5 in (34 cm). 1961-1964. Scarcity: 7. $325-$625.

(left) #5771 **Sonicon Rocket mini replica** with radar antenna. Windup, tin and plastic. 5 in (13 cm). 1997-1998. Scarcity: 2. $10-$20. (right) #3043b **Sonicon Rocket** with tin pilot, rotating antenna, lights, and whistle sound direction control. Lt blue/silver metal panel design. Battery operated, tin with plastic. 14 in (36 cm). 1964-1967. Picture illustrates original and 4-inch replica. Scarcity: 8. $350-$650.

#3241 **Space Pioneer** with pilot, sound, and flashing lights. Battery operated, mystery action, tin. 12 in (30 cm). 1962-1967. Scarcity: 6. *Courtesy of Cybertoyz.* $400-$650.

#3966 **Sonicon Space Rocket** with spaceman in cockpit, radar antenna, and whistle directional control. Battery operated, tin and plastic. 14 in (36 cm). 1971-1973. Scarcity: 6. $300-$500.

Space Ship Rocket. Friction, tin. 12.5 in (32 cm). 1950s. Scarcity: 4. *Courtesy of Justin Pinchot.* $1,600-$2,200.

194 Space

#3840 **Docking Apollo**. Capsule travels around lithographed track to dock with Lunar Module. Battery operated, tin. 18.5 in (47 cm). 1970-1971. Scarcity: 4. $200-$375.

Space Ship X-3 with sparking. Friction, tin. 8 in (20 cm). 1953. Scarcity: 6. *Courtesy of Justin Pinchot.* $400-$600.

Space Ship X-5. Friction, tin. 12 in (30 cm). 1952. Scarcity: 7. *Courtesy of Justin Pinchot.* $600-$800.

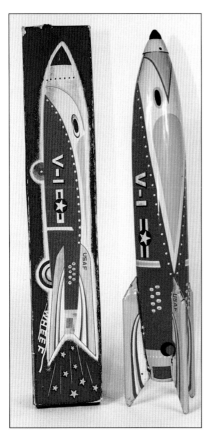

Space Ship V-1. Cragstan version of Space Ship Rocket. Friction, tin. 12.5 in (32 cm). 1950s. Scarcity: 3. $800-$1,400.

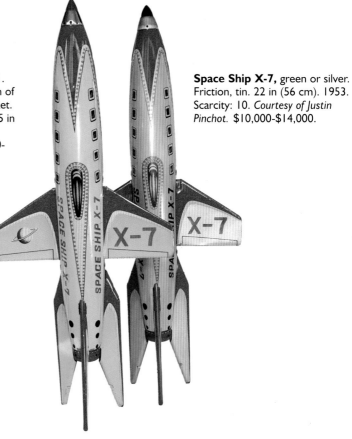

Space Ship X-7, green or silver. Friction, tin. 22 in (56 cm). 1953. Scarcity: 10. *Courtesy of Justin Pinchot.* $10,000-$14,000.

Space Ship X-2. Friction, tin. 7.5 in (19 cm). 1952. Scarcity: 5. $125-$250.

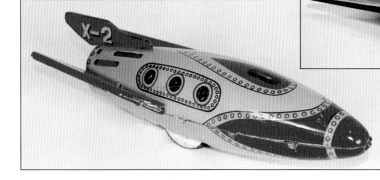

Space Ship X-7 Milky Way. Hand painted mock-up sample that was not produced with Milky Way. Friction, tin. 22 in (56 cm). 1953. Scarcity: 10.

Space Exploration and Travel 195

#1718A **Billy The Space Traveler** with tin astronaut riding on outside of "Billy" rocket. Friction, tin. 8 in (20 cm). 1957-1960. Scarcity: 8. $400-$800.

#1718B **Billy The Space Traveler** with tin astronaut riding on outside of "Billy" rocket. Friction, tin. 8 in (20 cm). 1959-1960. Scarcity: 8. $400-$800.

(left) #3791s **Space Patrol** hand painted mock-up sample for 3791. Friction, tin. 11 in (28 cm). 1969. Scarcity: 10. (right) **Space Ship X-8**. Hand painted mock-up sample. No known production. Friction, tin. 12.25 in (31 cm). 1953.

#1783 **Atomic Rocket with Prop**. "Atomic Rocket" with plastic circular propeller. Friction, tin with plastic. 7.5 in (19 cm). 1957-1958. Scarcity: 5. $100-$200.

#3122 **Mars Rocket** with lighted cockpit and tin pilot with moving arms. Battery operated, tin. 14.5 in (37 cm). 1960-1962. Scarcity: 8. $300-$500.

Moon Rocket with one-dimensional lithographed pilot. Friction, tin. 7 in (18 cm). 1952. Scarcity: 6. $200-$400.

196 Space

#1469b **Rocket Racer No.3** with tin or vinyl pilot and detonation. Friction, tin. 7 in (18 cm). 1960-1969. Reintroduced as number #3983 in 1972. Scarcity: 3. $75-$150.

#1556 **Fire Bird with Detonation**. No. 308 rocket racer with tin pilot and detonation sound. Friction, tin. 12.5 in (32 cm). 1955-1960. Scarcity: 5. *Courtesy of Smith House Toys.* $100-$200.

#1885 **Space Rocket No.9 with Detonation** with vinyl head pilot and clear windshield. No.9 on front of rocket. Friction, tin. 13 in (33 cm). 1959-1970. Model changed to #3961 in 1971. Scarcity: 3. $125-$250.

Rocket X-6 with one-dimensional lithographed pilot. Friction, tin. 3.75 in (10 cm). 1952. Scarcity: 6. *Courtesy of Justin Pinchot.* $150-$250.

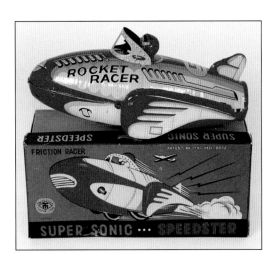

#1469a **Rocket Racer No.7** with tin pilot and detonation. Friction, tin. 7 in (18 cm). 1953-1959. Scarcity: 7. $150-$300.

Super Sonic Speedster Rocket Racer No.5. No. 5 with tin pilot and detonation sound. Friction, tin. 7 in (18 cm). 1952. Scarcity: 6. $200-$350.

#5773 **Rocket Racer mini replica**. Windup, tin. 1997-1998. Scarcity: 1. $10-$20.

Space Exploration and Travel 197

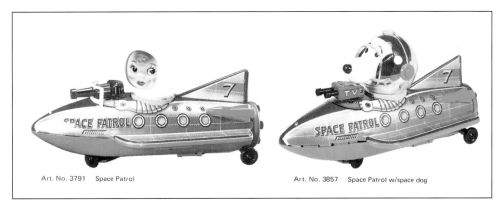

(left) #3791 **Space Patrol** with oversize vinyl space pilot and gun with light and sound. Battery operated, mystery action, tin with vinyl. 11 in (28 cm). 1970-1972. Scarcity: 4. $75-$125. (right) #3857 **Space Patrol w/Space Dog** with oversized "Snoopy"-like dog in plastic helmet, moving TV camera, and sound. Battery operated, mystery action, tin and plastic. 11 in (28 cm). 1971-1974. Scarcity: 4. $125-$250.

Flying Saucer Z-101 with sparking and astronauts lithographed in windows. Friction, tin. 7 in (18 cm). 1950s. Scarcity: 6. *Courtesy of Stan Luksenburg.* $200-$375.

(left) #3961 **Rocket Racer** "Space Rocket" with vinyl-head boy and detonation sound. Battery operated, tin with vinyl. 13 in (33 cm). 1971-1973. Scarcity: 5. $200-$300. (right) #3983 **Rocket Racer No.3** with sound and vinyl pilot. Friction, tin. 6.75 in (17 cm). 1972-1973. Scarcity: 3. $75-$150.

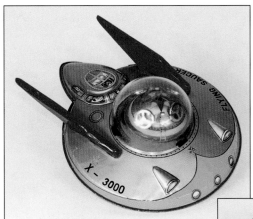

#1862 **Flying Saucer X-3000** with rotating circular tank inside plastic dome. Friction, tin with plastic. 6.5 in (17 cm). 1958-1960. Scarcity: 10. *Courtesy of Stan Luksenburg.* $400-$750.

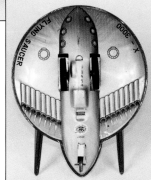

Flying Saucer Z-106 with sparking and lithographed crew in windows. Friction, tin. 5.5 in (14 cm). 1950s. Scarcity: 6. $250-$400.

198 Space

#3242A **Space Giant Explorer** with sound and flashing lights. Battery operated, mystery action, tin. 12 in (30 cm). 1962-1967. Scarcity: 7. *Courtesy of Cybertoyz.* $600-$1,000.

#3323 **Flying Saucer X-3** with tin or vinyl head pilot and sparking in red dome. Friction, tin. 5.5 in (14 cm). 1964. Scarcity: 6. *Courtesy of Stan Luksenburg.* $300-$600.

#3242B **Space Ship 1** with sound and flashing lights. Battery operated, mystery action, tin. 12 in (30 cm). 1962-1967. Scarcity: 7. *Courtesy of Cybertoyz.* $400-$800.

#3246 **Capsule 6** with tin pilot, sound, and flashing lights. Battery operated, mystery action, tin. 10.5 in (27 cm). 1962-1967. Scarcity: 9. *Courtesy of Smith House Toys.* $400-$800.

#3217 **Satellite X-107 with Astronaut in Orbit** with air floating Styrofoam astronauts that follow satellite and interior figure silhouettes on screen. Battery operated, mystery action, tin. 8 in (20 cm). 1962-1967. Picture illustrates Cragstan version. Scarcity: 6. $175-$350.

Space Exploration and Travel   199

#3052 **Space Patrol Trio** X-15, X-16, X-17. *Fineas J. Whoopie Collection.*

#3052a **Space Patrol with Floating Satellite X-15**. X-15 with air floating Styrofoam ball. Battery operated, non-fall, tin. 8 in (20 cm). 1960-1962. Scarcity: 6. *Fineas J. Whoopie Collection.* $150-$350.

#3052b **Space Patrol with Floating Satellite X-16**. X-16 with air floating Styrofoam ball. Battery operated, non-fall, tin. 8 in (20 cm). 1960. Scarcity: 6. *Fineas J. Whoopie Collection.* $150-$350.

#3862 **Space Surveyor X-12**. Satellite-like spaceship rotates with flashing light and rotating lithographed interior images. Battery operated, mystery action, tin. 7.75 in (20 cm). 1971. Scarcity: 8. *Courtesy of Cybertoyz.* $300-$600.

#3117 **Space Patrol with Floating Satellite X-17** with air floating ball. Battery operated, mystery action, tin. 8 in (20 cm). 1960-1962. Scarcity: 6. *Fineas J. Whoopie Collection.* $150-$350.

200 Space

#2664 **UFO** with spinner in three color variations. Friction, tin and plastic. 3 in (8 cm). 1979-1980s. Scarcity: 4. $15-$25.

#2670 **UFO** with roulette game in dome in two color variations. Friction, tin and plastic. 5 in (13 cm). 1979-1980s. Scarcity: 4. $25-$35.

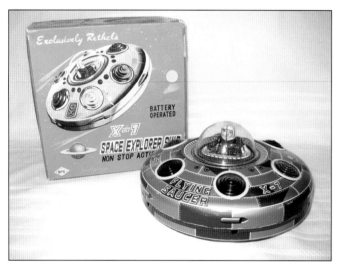

#1994 **Flying Saucer X-7** with tin pilot and multi-color lights outside dome. Light blue/dark blue metal paneled lithography. Battery operated, mystery action, tin. 8 in (20 cm). 1960-1970. Vinyl head on pilot around 1969. Scarcity: 4. $150-$250.

#3150 **Planet Explorer X-80**. X-80 saucer space ship with lighted plastic dome cockpit. Battery operated, non-stop, tin with plastic. 8 in (20 cm). 1961-1967. Scarcity: 5. $150-$300.

#3735 **Space Ship X-5**. Blue saucer with astronaut pilot, antenna, and flashing interior lights. Battery operated, mystery action, tin and plastic. 8.25 in (21 cm). 1969-1974. Scarcity: 4. $75-$150.

#3678 **Space Explorer Ship X-8** with astronaut pilot, flashing lights, and sound. Battery operated, mystery action, tin with plastic. 10.25 in (26 cm). 1968-1970. Scarcity: 5. $125-$225.

Space Exploration and Travel   201

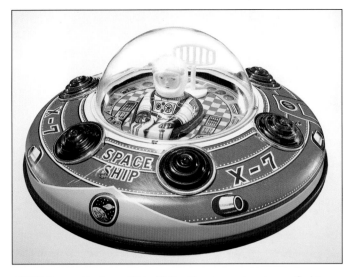

#3883 **Space Explorer Ship X-7** with astronaut, antenna, flashing perimeter lights, and sound. Battery operated, mystery action, tin with plastic. 8 in (20 cm). 1971-1974. Scarcity: 4. $75-$150.

#4316 **Space Explorer Ship X-7**. Blue saucer with astronaut, antenna, flashing perimeter lights, and sound. Battery operated, mystery action, tin and plastic. 8 in (20 cm). 1975-1980s. Scarcity: 3. $75-$150.

#4085 **Airport Saucer**. Spinning saucer with lighted tower and two jet planes circling within dome. Battery operated, mystery action, tin with plastic. 8.5 in (22 cm). 1972-1980. Scarcity: 5. $125-$225.

#4317 **Space Ship X-5**. Green saucer with astronaut pilot, antenna, and flashing interior lights. Battery operated, mystery action, tin and plastic. 8.25 in (21 cm). 1975-1980s. Scarcity: 3. $75-$150.

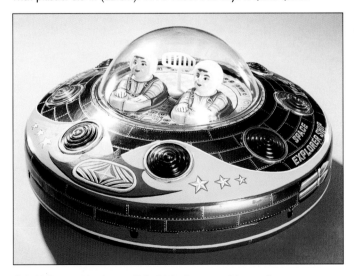

#4143 **Space Explorer Ship X-3**. Green or blue with sound, moving antenna, flashing lights outside dome, and twin astronauts. Battery operated, mystery action, tin with plastic. 10 in (25 cm). 1973-1974. Scarcity: 5. $75-$150.

#4595 **X-5 Space Ship (Silver)** with revolving body, pilot, antenna, and interior flashing lights. Battery operated, mystery action, tin and plastic. 8.25 in (21 cm). 1977-1980s. Scarcity: 3. $50-$100.

202 Space

#4670 **UFO X-05 Space Ship** with moving plastic pilot, antenna, and flashing lights outside cockpit. Battery operated, mystery action, tin with plastic. 7.5 in (19 cm). 1978-1980s. Scarcity: 3. $50-$100.

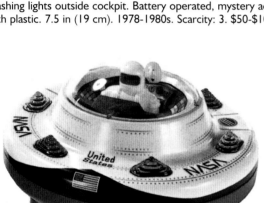

#4722 **Space Ranger No. 3** with moving plastic pilot, antenna, and rectangular flashing lights outside cockpit. Battery operated, tin with plastic. 7.5 in (19 cm). 1979-1980s. Scarcity: 5. $100-$200.

#4723 **Space Ranger No. 5** with moving cockpit and antenna and flashing lights in cockpit. Battery operated, tin with plastic. 8 in (20 cm). 1979-1980s. Scarcity: 4. $75-$150.

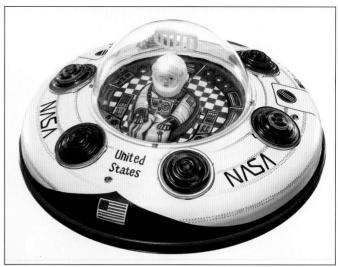

#4724 **Space Ranger No. 7** with moving cockpit and antenna and flashing lights on perimeter. Battery operated, tin with plastic. 8 in (20 cm). 1979-1980s. Scarcity: 4. $75-$150.

#4750 **Radicon Space Saucer** with single channel controller and space saucer. R/C battery operated, plastic. 8 in (20 cm). 1979-1980s. Scarcity: 5. $75-$150.

#4804 **Space Ship M-50** with flashing lights and robot like pilot. Battery operated, tin with plastic. 8 in (20 cm). 1980s. Scarcity: 5. $50-$100.

Space Exploration and Travel   203

#4953 **X-8 Space Ship**. Battery operated, tin with plastic. 8 in (20 cm). 1982-1980s. Scarcity: 2. $40-$75.

#4662 **Space Shuttle**. NASA shuttle with flashing light and sound. Battery operated, plastic. 10 in (25 cm). 1978-1980s. Scarcity: 4. $30-$50.

#8000 **Mars Explorer** with alien pilot and multi-color lights outside dome. Silver color. Battery operated, mystery action, tin and plastic. 8.25 in (21 cm). 1997. Imported by Rocket-USA. Scarcity: 2. $20-$40.

#4732 **Radicon Space Shuttle** with 2 channel radio control and sound. Battery operated, tin and plastic. 10 in (25 cm). 1979-1980. Scarcity: 4. $50-$100.

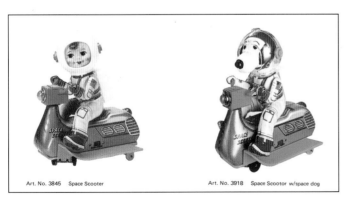

(left) #3845 **Space Scooter** with vinyl headed boy astronaut and blinking lights. Battery operated, mystery action, tin and plastic. 9 in (23 cm). 1970-1974. Scarcity: 4. $100-$200. (right) #3918 **Space Scooter w/Space Dog**. Snoopy-like dog riding scooter with flashing lights. Battery operated, mystery action, tin and plastic. 9.5 in (24 cm). 1971-1973. Scarcity: 5. $200-$325.

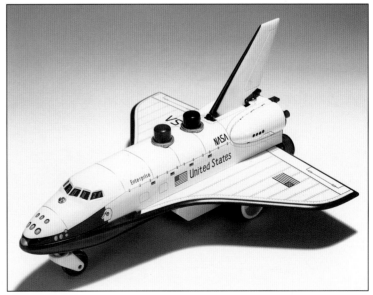

#4852 **Space Shuttle with Light** with dual flashing lights and space sound. Battery operated, non-fall, plastic. 10 in (25 cm). 1981-1980s. Scarcity: 2. $25-$45.

204 Space

#1856 **Radar Tank** with light and rotating plastic antenna. Battery operated, tin. 8 in (20 cm). 1958-1960. Scarcity: 5. *Courtesy of Cybertoyz.* $175-$350.

#1870 **Space Tank S-100** with large antenna and lithographed space pilots on windows. Friction, tin. 4 in (10 cm). 1958. Scarcity: 5. *Courtesy of Stan Luksenburg.* $100-$175.

#3386 **Planet Tank** with sound, antenna, and lighted dome. Battery operated, non-fall, tin with plastic. 8 in (20 cm). 1966-1967. Scarcity: 5. $175-$350.

#3659 **Lunar Expedition w/ Lever** with astronaut driver, light, and antenna lever actuated control. Battery operated, tin with plastic. 13.5 in (34 cm). 1968-1972. Scarcity: 5. $225-$450.

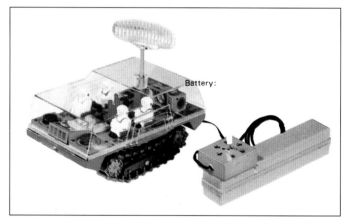

#3844 **Lunar Transport** with lever R/C control, four space figures, radar antenna, and blinking lights. R/C battery operated, tin with plastic. 8.5 in (22 cm). 1970-1974. Scarcity: 4. $75-$175.

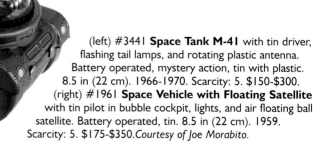

(left) #3441 **Space Tank M-41** with tin driver, flashing tail lamps, and rotating plastic antenna. Battery operated, mystery action, tin with plastic. 8.5 in (22 cm). 1966-1970. Scarcity: 5. $150-$300.
(right) #1961 **Space Vehicle with Floating Satellite** with tin pilot in bubble cockpit, lights, and air floating ball satellite. Battery operated, tin. 8.5 in (22 cm). 1959. Scarcity: 5. $175-$350. *Courtesy of Joe Morabito.*

Space Exploration and Travel 205

#4327 **MS-27 Missile Tank** with flying winged missile launched by remote control. R/C battery operated, plastic with tin. 12 in (30 cm). 1975-1976. Scarcity: 7. $200-$400.

#3660 **Radicon Space Pathfinder** with astronaut driver, light, antenna, and push button radio control. R/C battery operated, tin with plastic. 13.5 in (34 cm). 1968-1971. Scarcity: 6. $250-$450.

#4739 **Radicon Space Tank** with single channel radio control. R/C battery operated, plastic. 10 in (25 cm). 1979. Scarcity: 5. $40-$80.

#3861 **Space Guard MS-61 Tank** with forward/reverse, turning and shooting control fires missiles. Astronauts lithographed on front. R/C battery operated, tin with plastic. 8.75 in (22 cm). 1971-1973. Scarcity: 6. *Courtesy of Cybertoyz.* $250-$450.

206 Space

#3147 **Space Surveillant X-07** with tin pilot, sound, and flashing lights (vinyl pilot introduced 1969). Battery operated, non-fall, tin. 8.75 in (22 cm). 1961-1972. Scarcity: 4. *Fineas J. Whoopie Collection.* $125-$250.

#3175 **Space Survey X-09** with tin pilot, sound, and flashing lights (vinyl pilot introduced 1969). Battery operated, mystery action, tin. 8.75 in (22 cm). 1962-1970. Also sold as Cragstan Satellite Outer Space Survey Ship. Scarcity: 5. *Courtesy of Cybertoyz.* $150-$275.

#1789 **Flying Saucer with Space Patrol** with circular flying propellers. Friction, tin with plastic. 6.5 in (17 cm). 1957-1958. Scarcity: 8. $200-$300.

#1807 **Atomic Rocket with Prop**. X201 rocket with two plastic circular propellers. Friction, tin with plastic. 7.5 in (19 cm). 1958-1961. Scarcity: 5. $100-$200.

#1849 **Shooting Set No.2** with dart shooting space pistol and target board with satellite space graphics. Tin. 7 in (18 cm). 1958-1960. Scarcity: 5. $50-$100.

#1865 **Explorer Prop**. Pulling string in hand held grip causes circular flying propeller to spin upwards. Plastic. 7 in (18 cm). 1958. Scarcity: 4. $25-$50.

#1873 **Rocket Target Game** with dart gun. Hitting target launches rocket. Tin and fiber board. 9.5 in (24 cm). 1958-1960. Scarcity: 6. $200-$350.

#1878 **Space Express** with space ship cable cars traveling toward space station and back. Battery operated, tin. 9 in (23 cm). 1958. Scarcity: 9. $300-$600.

#1889 **Space Rocket and Tank** set with rocket X-200 and tank S-100 in common box. Friction, tin. 9 in (23 cm). 1958. Scarcity: 10. $400-$700.

#1912 **Space Tank M-18** with tin driver with moving arm, rotating plastic antenna, and rear springs. Battery operated, non-stop, tin with plastic. 8.5 in (22 cm). 1958-1964. Scarcity: 6. $200-$400.

#1968 **Space Vehicle with Prop**. Space Tank X-8 with circular flying plastic propeller. Friction, tin. 6 in (15 cm). 1960. Scarcity: 5. $125-$200.

#1994A **Space Explorer Ship X-7**. Blue saucer with astronaut, flashing perimeter lights, and sound. Battery operated, mystery action, tin. 8 in (20 cm). 1971. Scarcity: 5. $125-$250.

#3053 **Space Survey with Floating Satellite**. Space tank with air floating ball above. Battery operated, non-stop, tin. 7.5 in (19 cm). 1960-1962. Scarcity: 5. $150-$275.

#3160 **Acrobatic Rocket**. Rocket car rides inside rail of loop on stand. Friction, tin. 1961-1962. Scarcity: 7. $200-$300.

#3247a **Friendship Space Survey X-61 w/Floating Astronaut** with tin pilot, sound, and rotating (floating) tin astronaut. Battery operated, mystery action, tin. 10 in (25 cm). 1962-1964. Scarcity: 8. $300-$600.

#3841 **Space Capsule with Floating Astronaut** with astronaut that floats on air over capsule with blinking booster. Battery operated, mystery action, tin with plastic. 10 in (25 cm). 1970-1974. Imported by Illfelder. Scarcity: 5. $100-$200.

#3853 **X-11 Flying Saucer** with blinking lights and three astronauts, one of which floats. Battery operated, mystery action, tin with plastic. 7.75 in (20 cm). 1970-1971. Scarcity: 6. $150-$300.

#4854 **Space Shuttle**. NASA shuttle Enterprise with 3 channel, 4 function Radicon control. Battery operated, tin and plastic. 10 in (25 cm). 1981-1980s. Scarcity: 5. $75-$125.

#1864 **Space Age**. Two space rockets circle the globe while satellite rotates on wire overhead. Windup, tin. 6 in (15 cm). 1958. Scarcity: 8. *Courtesy of Justin Pinchot.* $400-$600.

Space Exploration and Travel – Not Pictured

Space Rocket Car 7 with vinyl space pilot and jet noise. Friction, tin. 12.75 in (32 cm). 1960s. Scarcity: 8. $150-$275.

Space Ship X-4 with windows cut out of body and celluloid tip. Friction, tin with celluloid. 6 in (15 cm). 1953. Scarcity: 7. $500-$750.

#1756 **Billy's Flying Saucer Z-112** with tin astronaut riding on outside of "Billy" saucer Z-112. Friction, tin. 7 in (18 cm). 1957-1958. Scarcity: 9. $750-$1,600.

Plasman

(top) #4427 **Plasman No.1**. Prime force Z Machine motor robot used to power system toys. Windup, plastic. 4 in (10 cm). 1976-1980s. (bottom) #4428 **Plasman No.2**. Prime force motor robot with power arms. Windup, plastic. 4 in (10 cm). 1976-1980s.

#4429-30-31 **Plasman Components** to be added to Plasman 1. Windup, plastic. 1976-1980s.

#4465-66-67 **Deluxe Plasman No. 2 with Components**. Windup, plastic. 1977-1980s.

#4468 **Deluxe Plasman No. 2 with Running Machine**. Windup, plastic. 1977-1978.

208 Space

#4538 **Plasman B Set** with Gear Machine. Windup, plastic. 1977-1980s.

#4537 **Plasman A Set** with Plasman boxers. Windup, plastic. 1977-1980s.

#4504 **Zel Robot Set A-E**. Metallic silver Plasman packaged with different accessories in sets A-E. Windup, plastic. 1979-1980s. Set E shown.

Plasman – Not Pictured

#4432 **Plasman Missile Launcher** to be added to Plasman 2. Windup, plastic. 1976-1978.

#4433 **Plasman Power Crane** to be added to Plasman 1. Windup, plastic. 1976-1980s.

#4435 **Plasman Running Machine** to be added to Plasman 2. Windup, plastic. 1976-1979.

#4469 **Deluxe Plasman No. 2 with Trailers**. Windup, plastic. 10 in (25 cm). 1977-1978.

#4477 **Plasman No.2 Set** with running machine, trailers, drill and shovel. Windup, plastic. 1977-1980s.

#4495 **Plasman No 2 w/ missile launcher**. Windup, plastic. 6 in (15 cm). 1977-1978.

#4539 **Plasman C Set** with Magnet Car. Windup, plastic. 1977-1980s.

#4540 **Plasman D Set** with Missile Vehicle. Windup, plastic. 1977-1980s.

Trains

Trains have always been popular with children as they represented one of the major transportation methods throughout the world. Trains were sold individually, as sets, in dioramas, and on platform bases. However, Masudaya's battery operated locomotive was one of the greatest merchandising successes in their history.

With the help of the realistic whistling sound described in the historical overview, the locomotives hit the market and sold unbelievably well. The first one with the famous hooting sound mechanism (No.3140 *Overland Express*) came out in 1961 and the last model was produced in 1981. During these twenty years, Masudaya produced more than two hundred different types, models, colors, and finishes of toy trains. These included items such as the western type steam engine with a large cowcatcher, streamlined modern locomotives with diesel engines, and comical locomotives with animals or kiddy characters. Altogether, over ten million locomotives were produced and shipped out all over the world. At least 50% of these products were shipped to the United States.

Today, it is difficult to document all of the variations. Many customers wanted their own variation made by changing designs and colors, but only starting with a relatively small quantity such as 300 dozen. Most of these items were discontinued when the inventory was depleted. In addition, these items were not normally included in catalogs unless there was inventory available at the time of catalog editing. Consequently, there were many un-cataloged items that cannot be documented in this way.

Animal Trains

#3286 **Animal Choo Choo Locomotive** with sparking and pulling car with lithographed animals in windows. Friction, tin. 14 in (36 cm). 1964-1966. Scarcity: 5. $40-$75.

#3653 **Elephant Express** with elephant face, animal graphics, whistling, smoke, and animal driver. Battery operated, mystery action, tin with vinyl. 17 in (43 cm). 1968-1973. Scarcity: 3. $40-$60.

New trains, 1970.

210 Trains

#3770 **Puppy-Band Loco** with music, loco drum face, and puppy in cab with cymbals. Battery operated, mystery action, tin and plastic. 9.75 in (25 cm). 1969-1972. Scarcity: 4. $40-$60.

#4188 **Panda Family Musician Locomotive** with mother panda playing cymbals and baby panda shaking on boiler. Battery operated, mystery action, tin and plastic. 10.5 in (27 cm). 1973-1975. Scarcity: 3. $40-$60.

Art. No. 3928

Art. No. 3833

(left) #3928 **Puppy Musician Locomotive** with puppy pulling bell, clanging cymbal, and drum beat on engine front. Battery operated, mystery action, tin with plastic. 10.5 in (27 cm). 1971-1973. Scarcity: 3. $40-$60. (right) #3833 **Elephant Musician Loco** with bell ringing and cymbal playing elephant driver. Battery operated, mystery action, plastic with tin. 10.5 in (27 cm). 1970-1973. Scarcity: 3. $25-$50.

#4828 **Panda Engineer Locomotive** with panda engineer, sound, and light. Battery operated, mystery action, plastic. 6.25 in (16 cm). 1981-1980s. Scarcity: 2. $20-$35.

#4194 **Elephant Family Musician Locomotive** with mother elephant playing cymbals and baby elephant shaking. Battery operated, mystery action, tin and plastic. 10.5 in (27 cm). 1974-1976. Scarcity: 3. $30-$50.

#4354 **Seesaw Express**. Animal train with seesaw rocking action and whistle. Battery operated, mystery action, tin with plastic. 10.5 in (27 cm). 1975-1977. Scarcity: 4. $40-$75.

Animal Trains – Not Pictured

Animal Express with single pantograph. Friction, tin. 17 in (43 cm). 1950s. Scarcity: 7. $250-$450.

#1908 **Animal Train** engine with face, pulls three articulated and connected cars illustrated with animal graphics. Battery operated, non-stop, tin. 18 in (46 cm). 1958. Scarcity: 3. $50-$100.

#3388 **Tinkling Locomotive** with lithographed monkey engineer. Friction, tin. 7.5 in (19 cm). 1965. Scarcity: 2. $20-$40.

Diesel Trains

#4343a **Diesel Locomotive** with whistle, sound, and headlight. Battery operated, mystery action, tin with plastic. 12 in (30 cm). 1975. Scarcity: 7. $75-$125.

#4343b **Lehigh Valley Diesel Locomotive** with headlight, sound, and whistle. Photo shows "Lehigh Vallet." Battery operated, mystery action, tin with plastic. 12 in (30 cm). 1975. Scarcity: 10. No price found.

#4190 **Santa Fe Diesel Locomotive with smoke** with engine sound, whistle, headlight, and puffing smoke. Battery operated, mystery action, tin with plastic. 15.75 in (40 cm). 1973-1976. Scarcity: 3. $50-$100.

#4290 **Union Pacific Diesel Locomotive w/ Smoke** with whistle, sound, flashing headlight, and smoke. Battery operated, mystery action, tin with plastic. 15.75 in (40 cm). 1974-1976. Scarcity: 4. $50-$100.

212 Trains

#4370 **Santa Fe Diesel Locomotive** with whistle, sound, and flashing headlight. Battery operated, mystery action, tin with plastic. 17.25 in (44 cm). 1975-1980s. Scarcity: 3. $75-$125.

#4378s **DB Diesel Locomotive**. Deutsche Bahn diesel sample. Battery operated, mystery action, tin with plastic. 15.75 in (40 cm). 1975. Scarcity: 10.

#4378 **DB Diesel Locomotive**. Deutsche Bahn diesel with whistle, sound, and flashing headlight. Battery operated, mystery action, tin with plastic. 15.75 in (40 cm). 1975. Scarcity: 7. $125-$200.

#4380 **New Diesel Locomotive Blue** with whistle, sound, and flashing headlight. Battery operated, mystery action, tin with plastic. 17.25 in (44 cm). 1975-1980s. Scarcity: 2. $50-$100.

#4378B **TEE Diesel Locomotive**. Trans Europe Express diesel with whistle, sound, and flashing headlight. Battery operated, mystery action, tin with plastic. 15.75 in (40 cm). 1975. Scarcity: 8. $150-$250.

#4390 **New Santa Fe Diesel Locomotive - Yellow** with whistle, sound, and flashing headlight. Battery operated, mystery action, tin with plastic. 17.25 in (44 cm). 1976. Scarcity: 7. $100-$150.

Diesel Trains 213

#4480 **Mighty Diesel Locomotive,** European style with whistle and sound. Battery operated, mystery action, tin with plastic. 11.5 in (29 cm). 1976-1980s. Scarcity: 3. $40-$75.

#1681 **Electric Locomotive No.3.** Silver Arrow electric car with twin pantographs. Friction, tin. 16.5 in (42 cm). 1956-1960. Scarcity: 5. $200-$350.

#4690 **Mighty Diesel Locomotive w/ Light** with flashing top light and whistle and engine sound. Battery operated, mystery action, plastic with tin. 11.5 in (29 cm). 1978-1980s. Scarcity: 4. $40-$75.

Diesel Trains – Not Pictured

#4500 **New Diesel Locomotive Red** with whistle, sound, and flashing headlight. Battery operated, mystery action, tin with plastic. 17.25 in (44 cm). 1976-1977. Scarcity: 5. $75-$125.

#3431 **Stream-Liner "Glory"** with sound, red/green headlight, and lithographed passengers. Battery operated, mystery action, tin. 19 in (48 cm). 1966-1969. Scarcity: 5. $75-$125.

Electric Trains

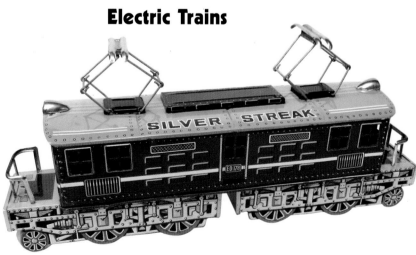

Electric Car Silver Streak with twin pantographs. Friction, tin. 17.5 in (44 cm). 1950s. Scarcity: 6. *Courtesy of Bill & Stevie Weart.* $250-$400.

#3538 **Radicon Dream Super Express.** Bullet train with lithographed passengers, antenna, and push button radio control. R/C battery operated, tin with plastic. 19 in (48 cm). 1967-1971. Scarcity: 6. $150-$250.

Trains

#4338 **Electromotive** with pantograph, lithographed passengers, and engine sound. Battery operated, mystery action, tin with plastic. 11.75 in (30 cm). 1975-1977. Scarcity: 6. $100-$175.

Platform Base Trains

New Turntable Tramway. Trolley turns around at each end of track with city scene backdrop. Windup, tin. 1930s. Scarcity: 8. *Courtesy of Ray Rohr*. $1,000-$1,800.

Happy Express with train circling on scenic round platform containing building, tunnels, and bridge. Windup, tin. 8 in (20 cm). 1950s. Scarcity: 5. *Courtesy of Rex and Kathy Barrett*. $125-$200.

#4339 **Electromotive** with twin pantographs and engine sound. Battery operated, mystery action, tin with plastic. 13.25 in (34 cm). 1975-1978. Scarcity: 5. $100-$175.

Electric Trains – Not Pictured

#1665 **Electric Locomotive No.2** with twin pantographs. Friction, tin. 10 in (25 cm). 1956-1960. Scarcity: 4. $150-$250.

#1686 **Electric Car No.2** with single pantograph. Friction, tin. 12 in (30 cm). 1957-1960. Scarcity: 5. $200-$300.

#1778 **Electric Car Set No. 400**. Electric Car 400 with pantograph travels 2-rail track. Battery operated, tin. 12.5 in (32 cm). 1957-1958. Scarcity: 6. $200-$300.

#1829 **Electric Locomotive No.4** with twin pantographs and lithographed engineer in window. Friction, tin. 12.5 in (32 cm). 1958-1960. Scarcity: 5. $200-$300.

#3279 **Super Express Train**. 3 car bullet train with pantographs. Friction, tin. 1963-1964. Scarcity: 4. $50-$100.

#3285 **Central Electric Car No.4**. Central Line with single pantograph. Friction, tin. 10.5 in (27 cm). 1964. Scarcity: 5. $150-$225.

Return Tram moves back and forth on 2 piece track platform. Windup, tin. 22 in (56 cm). 1958. Scarcity: 6. $150-$250.

Shunting Train with engine and freight car shunting back and forth on elevated platform. Windup, tin. 12 in (30 cm). 1950s. Scarcity: 5. $200-$350.

Steam Trains 215

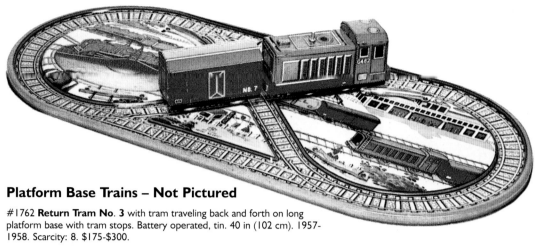

#3805 **Hard Working Diesel Locomotive**. Diesel shuttles train car around lithographed platform base. Battery operated, tin. 18.5 in (47 cm). 1970-1971. Reissue of #1814. Scarcity: 4. $100-$150.

Platform Base Trains – Not Pictured

#1762 **Return Tram No. 3** with tram traveling back and forth on long platform base with tram stops. Battery operated, tin. 40 in (102 cm). 1957-1958. Scarcity: 8. $175-$300.

#1763 **Return Bus** with bus traveling back and forth on platform base with bus stops. Windup, tin. 22 in (56 cm). 1957-1958. Scarcity: 6. $125-$200.

#1814 **Hard Working Diesel Locomotive**. Diesel shuttles train car around lithographed platform base. Battery operated, tin. 18.5 in (47 cm). 1957-1962. Scarcity: 4. $125-$175.

#1841 **Mono-rail** travels on elevated wire ring. Includes car with lithographed passengers in windows. Windup, tin. 5 in (13 cm). 1958-1960. Scarcity: 7. $100-$200.

Steam Trains

A-101 Locomotive with detonation sound. Friction, tin. 7 in (18 cm). 1950s. Scarcity: 4. *Courtesy of Barbara Moran.* $60-$100.

Sparky The Locomotive. D1620 with sparking. Windup, tin. 5.5 in (14 cm). 1950s. Scarcity: 4. *Courtesy of Barbara Moran.* $50-$75.

Locomotive Press-a-Toy. Locomotive D1632 with press down action. Friction, tin. 6 in (15 cm). 1956. Scarcity: 5. $50-$100.

#1904B **Super Express Train** with engineer lithographed on window. Friction, tin. 19.5 in (50 cm). 1958. Scarcity: 4. $50-$100.

216   Trains

#1934 **Blow-up Ball Locomotive** with ball held up by blowing air from smokestack. Battery operated, mystery action, tin. 10 in (25 cm). 1959. Scarcity: 4. $75-$125.

#3191 **Hooting Locomotive** with hooting sound. Battery operated, mystery action, tin. 9.75 in (25 cm). 1962-1967. Imported by Rosko. Scarcity: 2. $50-$100.

#2169 **Whistling See-Thru Locomotive** with see-though body showing gears and direction set by position of driver. Windup, plastic. 6 in (15 cm). 1978-1979. Scarcity: 4. $30-$50.

(top) #3230 **Western Special Locomotive**. Western loco with whistle, sound, and flashing light. Battery operated, mystery action, tin. 15.25 in (39 cm). 1963-1973. Scarcity: 1. $25-$45. (bottom) #5769 **Western Special mini replica** of Western loco. Windup, tin. 1997-1998. Scarcity: 1. $10-$20.

#3232 **Golden Arrow Train**. All black with engineer silhouette in cab. Friction, tin. 14 in (36 cm). 1963. Scarcity: 5. *Courtesy of Barry Skelly.* $50-$75.

#3140 **Overland Express** with whistling sound and Indian Head logo. This toy was the first of the many big whistling locos produced by Masudaya. Battery operated, mystery action, tin. 16 in (41 cm). 1961-1967. Scarcity: 2. $50-$100.

Steam Trains 217

#3291 **Great Western Locomotive** with whistle, headlight, and boiler light. Battery operated, mystery action, tin. 20.75 in (53 cm). 1963-1967. Scarcity: 4. $50-$100.

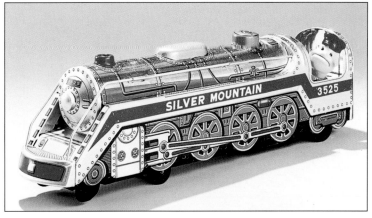

#3525 **Silver Mountain Locomotive** with whistle sound, flashing engine and shaking engineer. Battery operated, mystery action, tin with vinyl. 16 in (41 cm). 1967-1973. Scarcity: 1. $25-$45.

#3305 **Silver Streak Express** with whistle, smoke, and headlight. Battery operated, mystery action, tin. 16 in (41 cm). 1964-1972. Scarcity: 1. $30-$50.

#3600 **Silver Frontier Special Loco** with whistle and sound. Battery operated, mystery action, tin with plastic. 17.5 in (44 cm). 1968-1971. Scarcity: 3. $30-$50.

#3408 **Tinkling Locomotive** with tinkling sound and shaking vinyl driver. Battery operated, mystery action, tin with vinyl. 9 in (23 cm). 1965-1967. Scarcity: 5. $30-$50.

#3634 **Tiny Express Locomotive** with forward and reverse control. R/C battery operated, tin. 7 in (18 cm). 1968-1970. Scarcity: 5. $40-$75.

#3635 **Giant Western Special Express** with bell, whistle, and flashing smoke stack. Battery operated, mystery action, tin with plastic. 18.5 in (47 cm). 1968-1973. Scarcity: 2. $50-$100.

#3463 **Toy Town Express w/Fire Man** with vinyl head engineer, flashing stack, and shaking engine. Battery operated, mystery action, tin with vinyl. 11 in (28 cm). 1966-1967. Scarcity: 5. $50-$75.

218   Trains

#3648 **Overland Express (small size)** with whistling sound and lights. Battery operated, mystery action, tin. 11.25 in (29 cm). 1968-1971. Scarcity: 3. $40-$60.

#3649 **Silver Western Special Locomotive** with whistle and sound. Battery operated, mystery action, tin with plastic. 14.5 in (37 cm). 1968-1972. Scarcity: 2. $40-$60.

#3661 **Silver Arrow Locomotive** with flashing headlight and sound. Battery operated, mystery action, tin. 11.25 in (29 cm). 1968-1972. Scarcity: 2. $30-$50.

#3671 **Mountain Special Express** with engine face, flashing light and engine, sound, and shaking engineer driver. Battery operated, mystery action, tin with vinyl. 16 in (41 cm). 1968-1973. Scarcity: 2. $30-$50.

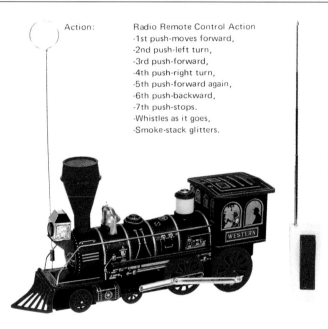

#3700 **Radicon Giant Western Loco** with whistle, antenna, and push button radio control. R/C battery operated, tin with plastic. 18.5 in (47 cm). 1968-1971. Scarcity: 4. $100-$175.

#3711 **Silver Overland Express** with whistle and lights. Battery operated, mystery action, tin with plastic. 11.25 in (29 cm). 1969-1971. Scarcity: 4. $25-$45.

Steam Trains 219

#3738 **Ball Puffing Locomotive** with cowboy musician lithography, bell sound, and ball blowing stack. Battery operated, mystery action, tin with plastic. 10.5 in (27 cm). 1969-1971. Scarcity: 5. $50-$75.

#3744 **Western Special Locomotive w/ lantern** with sound and swinging lantern. Battery operated, mystery action, tin with plastic. 14.5 in (37 cm). 1969-1973. Scarcity: 2. $40-$60.

#3739 **Hop Up Locomotive** with smoke, bell, light, and hop-up stack. Battery operated, mystery action, tin with plastic. 10.5 in (27 cm). 1969-1972. Scarcity: 4. $40-$75.

(left) #3745 **Special Giant Western Locomotive** with flashing smoke stack and push button bell, smoke and whistle. Can also be used as a push toy. Battery operated, tin with plastic. 18.5 in (47 cm). 1969-1974. Scarcity: 3. $50-$100. (right) #3960 **Bell Clanger Western Choo Choo** with arm pulling bell, whistle, clang-clang sound, and flashing light. Battery operated, mystery action, tin with plastic. 14.5 in (37 cm). 1971-1973. Scarcity: 4. $50-$100.

#3743 **Bell Ringer Choo Choo** with bell sound and flashing stack. Battery operated, mystery action, tin with plastic. 10 in (25 cm). 1969-1971. Scarcity: 4. $40-$75.

#3777 **Western Pioneer Locomotive** with bell sound. Battery operated, mystery action, tin with plastic. 10.5 in (27 cm). 1970-1972. Scarcity: 3. $30-$50.

220 Trains

#4018 **Continental Blue Locomotive**. European style locomotive with whistle, light, and sound. Battery operated, mystery action, tin with plastic. 12.5 in (32 cm). 1972-1976. Scarcity: 2. $40-$75.

#3829 **Western Choo Choo** with vinyl driver swinging lantern. Battery operated, mystery action, tin with plastic. 11.25 in (29 cm). 1970-1972. Scarcity: 4. $30-$50.

#3971 **Silver Mountain Flash Express** with moving engineer, flashing headlight, and sound. Battery operated, mystery action, tin with plastic. 16 in (41 cm). 1972-1976. Scarcity: 2. $30-$50.

#4027 **Giant Benkei Express** with tender, whistle, shaking boiler, rattling sound, light, and puffing smoke. Battery operated, mystery action, tin with plastic. 17 in (43 cm). 1972. Used on Hokkaido Coalmine & RR in Japan. Scarcity: 9. $125-$200.

#3980 **Continental Express** with whistle sound, puffing smoke, and flashing headlight. Battery operated, mystery action, tin with plastic base, stack and light. 15.5 in (39 cm). 1972-1980. Scarcity: 1. $40-$80.

Steam Trains 221

#4040 **Frontier Whistling Locomotive** with flashing smoke stack and remote control for whistle. Battery operated, mystery action, tin with plastic. 9.25 in (23 cm). 1972-1974. Scarcity: 4. $30-$50.

#4079 **Bell Ringer Choo Choo** with bell sound. Battery operated, mystery action, tin. 10.5 in (27 cm). 1972-1973. Scarcity: 4. $40-$60.

#4062 **Silver Mountain Express -A-** with flashing light, sound, and lithographed engineer in cab window. Battery operated, mystery action, tin with plastic. 16 in (41 cm). 1972-1975. Scarcity: 3. $30-$50.

Art. No. 4191　　Art. No. 4192　　Art. No. 4193　　Art. No. 4067
Piston Red Mountain Express　Piston Blue Mountain Express　Piston Green Mountain Express　Piston Silver Mountain Express

(left to right) #4191 **Piston Red Mountain Express**. Red Mountain Express with moving pistons and moving engineer. Battery operated, mystery action, tin with plastic. 16 in (41 cm). 1973-1980s. Scarcity: 1. $20-$40.
#4192 **Piston Blue Mountain Express**. Blue Mountain Express with moving pistons and moving engineer. Battery operated, mystery action, tin with plastic. 16 in (41 cm). 1973-1980s. Scarcity: 1. $20-$40.
#4193 **Piston Green Mountain Express**. Green Mountain Express with moving pistons and moving engineer. Battery operated, mystery action, tin with plastic. 16 in (41 cm). 1973-1980s. Scarcity: 1. $20-$40.
#4067 **Piston Silver Mountain Express** with moving pistons, sound, and moving engineer. Battery operated, mystery action, tin with plastic. 16 in (41 cm). 1972-1980s. Scarcity: 1. $20-$40.

222 Trains

#4080 **Union Pacific Express** with whistle, shaking boiler, rattling sound, light, and puffing smoke. Battery operated, mystery action, tin with plastic. 14 in (36 cm). 1972-1973. Japanese market version marked "Benkei". Scarcity: 6. $100-$175.

#4197 **Continental Gold Locomotive**. European style locomotive with whistle, light, and sound. Battery operated, mystery action, tin with plastic. 12.5 in (32 cm). 1973-1976. Scarcity: 3. $40-$75.

#4195 **Continental Red Locomotive**. European style locomotive with whistle, light, and sound. Battery operated, mystery action, tin with plastic. 12.5 in (32 cm). 1973-1976. Scarcity: 3. $40-$75.

#4230 **New Silver Mountain Express** with moving engineer, flashing engine, and whistling sound. Battery operated, mystery action, tin with plastic. 16 in (41 cm). 1974-1980s. Scarcity: 1. $20-$40.

#4196 **Continental Green Locomotive**. European style locomotive with whistle, light and sound. Battery operated, mystery action, tin with plastic. 12.5 in (32 cm). 1973-1978. Scarcity: 2. $40-$75.

#4287 **Overland Express** with whistling sound and flashing headlight. Battery operated, mystery action, tin and plastic. 11.5 in (29 cm). 1974-1976. Scarcity: 3. $40-$60.

Steam Trains   223

#4291 **Blue Express** with moving driver, whistling sound, and flashing headlight. Battery operated, mystery action, tin and plastic. 11.5 in (29 cm). 1974-1976. Scarcity: 3. $40-$60.

(left) #4552 **Western Special Locomotive** with whistle sound, red headlight, and vibrating smoke stack. Battery operated, mystery action, plastic with tin. 12.5 in (32 cm). 1977-1980s. Scarcity: 3. $40-$60. (right) #4335 **Pioneer Express Steam Locomotive** with sound and whistle. Battery operated, mystery action, tin with plastic. 15 in (38 cm). 1975-1978. Scarcity: 4. $40-$60.

#4292 **Red Express** with moving driver, whistling sound, and flashing headlight. Battery operated, mystery action, tin and plastic. 11.5 in (29 cm). 1974-1976. Scarcity: 3. $40-$60.

#4381 **New Silver Western Locomotive** with chrome body and whistle. Battery operated, mystery action, plastic. 9.5 in (24 cm). 1975. Scarcity: 6. $60-$100.

#4478a **Transcontinental Express Loco** with whistle sound. Battery operated, mystery action, tin with plastic. 9.5 in (24 cm). 1976-1980s. Scarcity: 2. $35-$60.

#4314 **Continental Express steam locomotive** with train sounds. Battery operated, mystery action, tin with plastic. 11 in (28 cm). 1975-1976. Scarcity: 5. $40-$80.

Trains

#4478b **Transcontinental Express Loco.** Nacionales de Mexico (Mexican National Railroad) version of Transcontinental Express Loco. Battery operated, mystery action, tin with plastic. 9.5 in (24 cm). 1976. Scarcity: 6. $60-$100.

#4488 **New Western Loco (Black)** with whistle sound. Battery operated, mystery action, plastic. 9.75 in (25 cm). 1976-1977. Scarcity: 4. $30-$50.

#4479 **Iron Horse Special Locomotive** with whistle, bell, and sound. Battery operated, mystery action, tin with plastic. 14 in (36 cm). 1976-1980s. Scarcity: 2. $40-$80.

#4492 **Mini Overland Locomotive** with whistle sound and flashing light. Battery operated, mystery action, plastic with tin. 6.25 in (16 cm). 1976-1980s. Scarcity: 2. $20-$30.

#4597 **Piston Locomotive** with see through lighted engine and moving pistons. Battery operated, mystery action, plastic with tin. 6.25 in (16 cm). 1977-1980s. Scarcity: 2. $20-$40.

Steam Trains 225

#4603 **Transcontinental Silver Express** with whistle sound. Battery operated, mystery action, plastic with tin. 9.5 in (24 cm). 1977-1980s. Scarcity: 2. $30-$50.

#4604 **Transcontinental Blue Express** with whistle sound. Battery operated, mystery action, plastic with tin. 9.5 in (24 cm). 1977-1980s. Scarcity: 2. $30-$50.

(left to right) #4606 **Fireball Red Express** with see-through lighted engine, sound, and vinyl head engineer. Battery operated, mystery action, tin with plastic. 16 in (41 cm). 1977-1980s. Scarcity: 2. $30-$50. #4607 **Fireball Blue Express** with see-through lighted engine, sound and vinyl head engineer. Battery operated, mystery action, tin with plastic. 16 in (41 cm). 1977-1980s. Scarcity: 2. $30-$50. #4608 **Fireball Green Express** with see-through lighted engine, sound and vinyl head engineer. Battery operated, mystery action, tin with plastic. 16 in (41 cm). 1977-1980s. Scarcity: 2. $30-$50.

#4623 **Western Special Locomotive w/ Light** with whistle sound, red headlight, red boiler light, and vibrating smoke stack. Battery operated, mystery action, plastic with tin. 12.5 in (32 cm). 1978-1980s. Scarcity: 2. $40-$60.

#4678 **See-Saw Express** with lithographed figures in windows, see-saw rocking, and whistle sound. Battery operated, mystery action, plastic with tin. 10.5 in (27 cm). 1978-1980s. Scarcity: 4. $40-$60.

(left to right) #4841 **Continental Express** with whistle sound and flashing headlight. Battery operated, mystery action, tin and plastic. 15.5 in (39 cm). 1981. Scarcity: 3. $40-$60. #4684 **Piston Gold -A- Express**. Gold with whistle sound, moving pistons and lithographed engineer. Battery operated, mystery action, tin with plastic. 16 in (41 cm). 1978-1980s. Scarcity: 3. $20-$40. #4683 **Piston Silver -A- Express**. Silver with whistle sound, moving pistons and lithographed engineer. Battery operated, mystery action, tin with plastic. 16 in (41 cm). 1978-1980s. Scarcity: 3. $20-$40.

#4610 **Green Streak Express w/ Smoke** with whistle sound and smoke. Battery operated, mystery action, tin. 16 in (41 cm). 1977-1979. Scarcity: 4. $40-$60.

Steam Trains – Not Pictured

#1874 Bubble Locomotive with bubble blowing stack. Battery operated, tin with plastic. 9.5 in (24 cm). 1958. Scarcity: 5. $60-$100.

#1904 Mountain Giants Express Train. Steam train #48200 with engineer lithographed on window. Friction, tin. 19.5 in (50 cm). 1958. Scarcity: 4. $50-$100.

#3006 Western Locomotive with lithographed western and cow figures and sparking. Friction, tin. 6.5 in (17 cm). 1960-1967. Scarcity: 1. $20-$30.

#3020 Old Fashioned Locomotive modeled after the first steam locos. Friction, tin. 10 in (25 cm). 1960. Scarcity: 5. $75-$125.

#3032 Non Stop Super Loco with Whistle. 4-8-4 locomotive with tender and whistle. Battery operated, non-stop, tin. 19.5 in (50 cm). 1960-1962. Scarcity: 4. $50-$100.

#3033 Comic Loco with face, plastic stack, and comic arms driving pistons. Battery operated, non-stop, tin and plastic. 12.75 in (32 cm). 1959-1962. Scarcity: 2. $50-$75.

#3088 Western Pioneer with whistle. Battery operated, mystery action, tin. 10 in (25 cm). 1960-1962. Scarcity: 4. $40-$60.

#3177 Smoky Locomotive No.3177 with smoke and whistle sound. Battery operated, mystery action, tin. 9.75 in (25 cm). 1961-1967. Scarcity: 2. $40-$75.

#3181 Whoo Whoo Locomotive No. 1968 with bell, whistling sound, and lithographed windows. Battery operated, mystery action, tin. 11 in (28 cm). 1962. Scarcity: 5. $60-$100.

#3185 Western Choo Choo with oversized tin engineer, smoke and sound. Battery operated, mystery action, tin. 11 in (28 cm). 1961-1964. Scarcity: 3. $75-$100.

#3187 Silver Arrow Locomotive. Friction, tin. 14 in (36 cm). 1962-1964. Scarcity: 4. $50-$100.

#3202 Golden Falcon Locomotive No. 6681 with smoke and whistle sound. Battery operated, mystery action, tin. 16 in (41 cm). 1963-1967. Scarcity: 4. $60-$100.

#3210 Trans Continental Express No. 6601 with whistle sound and headlight. Battery operated, mystery action, tin. 16 in (41 cm). 1965-1967. Scarcity: 5. $60-$100.

#3231 Dream Train. Friction, tin. 10 in (25 cm). 1964-1964. Scarcity: 5. $50-$75.

#3234 Western Arrow Locomotive with smoke and whistle sound. Battery operated, mystery action, tin. 10 in (25 cm). 1963-1967. Scarcity: 3. $40-$60.

#3235 Central Choo Choo Locomotive with sound. Battery operated, tin. 15 in (38 cm). 1962. Scarcity: 5. $50-$100.

#3236 Western Lite-up Loco. Western loco with oversized tin engineer, flashing light, and sound. Battery operated, mystery action, tin. 11 in (28 cm). 1962-1967. Scarcity: 2. $50-$100.

#3237 Western Ringing Loco – Buffalo. Buffalo loco with oversized tin engineer, smoke, and sound. Stops to ring bell. Battery operated, mystery action, tin. 11 in (28 cm). 1962-1967. Scarcity: 2. $50-$100.

#3250 Overland Express No.6686 streamline with whistle sound. Battery operated, mystery action, tin. 11.25 in (29 cm). 1963-1966. Scarcity: 3. $40-$75.

#3293 Western Special Locomotive with Smoke with flashing light, smoke, and whistling choo choo sounds. Battery operated, mystery action, tin. 14.5 in (37 cm). 1964-1973. Scarcity: 1. $25-$45.

#3318 World Express with sound. Friction, tin. 16 in (41 cm). 1964-1967. Scarcity: 5. $50-$75.

#3430 Mountain Express w/Fire Man with sound, flashing engine, and shaking vinyl head driver. Battery operated, mystery action, tin with vinyl. 16 in (41 cm). 1966-1968. Scarcity: 3. $30-$50.

#3446 Western Express. Central locomotive with sound and flashing light. Battery operated, mystery action, tin. 14.5 in (37 cm). 1967-1968. Scarcity: 5. $40-$75.

#3502 Mountain Express II with multi-color headlight, sound, and smoke. Battery operated, mystery action, tin. 16 in (41 cm). 1967-1969. Scarcity: 3. $30-$50.

#3506 Silver Dart Locomotive with whistle sound. Battery operated, non-fall, tin. 11.25 in (29 cm). 1967. Scarcity: 6. $30-$50.

#3652 Ring-Smoke Choo Choo with smoke rings. Battery operated, mystery action, tin with plastic. 10.5 in (27 cm). 1968-1971. Scarcity: 3. $30-$50.

#3673 Frontier Locomotive with vinyl engineer swinging lantern, whistle, and sound. Battery operated, mystery action, tin with plastic. 17.5 in (44 cm). 1968-1971. Scarcity: 3. $35-$75.

#3676 Casey Jones Train with whistle and sound. Battery operated, mystery action, tin with plastic. 10.5 in (27 cm). 1968-1969. Scarcity: 4. $40-$75.

#3836 Bell Ringer with Lantern with vinyl driver swinging lantern and ringing bell. Battery operated, mystery action, tin with plastic. 10 in (25 cm). 1970-1973. Scarcity: 3. $30-$50.

#3860 Golden Cloud Express with moving engineer, flashing headlight, and sound. Battery operated, mystery action, tin with plastic. 16 in (41 cm). 1972-1975. Scarcity: 3. $30-$50.

#4150 New Western Special Locomotive with whistling sound, flashing light, and shaking boiler head. Battery operated, mystery action, tin with plastic. 14.5 in (37 cm). 1973. Scarcity: 6. $40-$60.

#4270 Western Special Locomotive with whistling sound and flashing light. Battery operated, mystery action, tin with plastic. 14.5 in (37 cm). 1974-1976. Scarcity: 3. $30-$50.

#4281 Western Special Locomotive with Smoke with whistling sound and flashing light. Battery operated, mystery action, tin with plastic. 14.5 in (37 cm). 1974-1976. Scarcity: 3. $35-$60.

#4282 Western Special Locomotive with Headlight with whistling sound and flashing light. Battery operated, mystery action, tin with plastic. 14.5 in (37 cm). 1974-1975. Scarcity: 4. $50-$75.

#4507 Silver Western Loco with whistle sound. Black with red. Battery operated, mystery action, plastic. 9.75 in (25 cm). 1976-1977. Scarcity: 4. $30-$50.

#4551 Western Locomotive with whistle sound. Friction, plastic. 9.75 in (25 cm). 1977. Friction version of #4488. Scarcity: 4. $30-$50.

#4609 Blue Streak Express w/ Smoke with whistle sound and smoke. Battery operated, mystery action, tin. 16 in (41 cm). 1977-1979. Scarcity: 4. $40-$60.

#4620 Western Special Locomotive w/ Smoke with whistle sound, smoke, and vibrating smoke stack. Battery operated, mystery action, plastic with tin. 12.5 in (32 cm). 1978-1980s. Scarcity: 2. $40-$60.

#4807 Western Special Locomotive with Buttons with two cab buttons for sound and movement. Battery operated, tin with plastic. 12.5 in (32 cm). 1980. Scarcity: 4. $40-$75.

#4844 Push Button Western Loco with three pushbuttons on cab for stop, go, and go with whistle and light. Battery operated, tin and plastic. 12.5 in (32 cm). 1981. Scarcity: 3. $40-$75.

Trolleys

#3754 Tinkling Trolley with five lithographed windows, light, driver, and bell sound. Battery operated, mystery action, tin with plastic. 13 in (33 cm). 1969-1972. Scarcity: 4. $75-$150.

1776 USA Bicentennial Trains   227

#3354 **Tinkling Trolley**. Broadway Trolley with eight lithographed windows, light, driver, and bell sound. Battery operated, mystery action, tin with plastic. 13 in (33 cm). 1966-1967. Scarcity: 4. $100-$175.

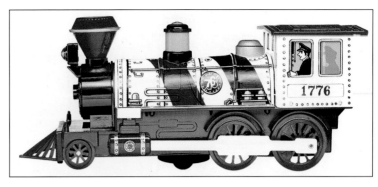

#4388 **Spirit of 1776 Western Locomotive** with US Bicentennial 1776 graphics. Battery operated, mystery action, tin with plastic. 15.25 in (39 cm). 1976. Scarcity: 4. $75-$125.

#3413 **R/C Tinkling Trolley** with lithographed passengers on windows and reversing trolley pole. R/C battery operated, tin. 9.75 in (25 cm). 1966-1967. Scarcity: 5. $75-$150.

#4388s **Spirit of 1776 European Locomotive**. Mock-up sample with US Bicentennial 1776 graphics. Battery operated, mystery action, tin with plastic. 12.5 in (32 cm). 1976. Scarcity: 10.

1776 USA Bicentennial Trains

In 1976, the United States celebrated its Bicentennial and the government sponsored a special Freedom Train named "Spirit of 1776" which traveled throughout the States. The chain store F. W. Woolworth showed interest in Masudaya's locomotive line and placed an order for a special version with the design similar to the actual original train. It is No.4406, Spirit of 1776 locomotive. Masudaya also produced a modern type diesel, western type with cowcatcher, and a European style locomotive with Bicentennial graphics.

#4406 **Spirit of 1776 Express** with US Bicentennial 1776 graphics, moving engineer, flashing engine, and whistling sound. Battery operated, mystery action, tin with plastic. 16 in (41 cm). 1976-1978. Scarcity: 3. $50-$100.

228　Trains

#4487 **Spirit of 1776 Diesel Locomotive** with US Bicentennial 1776 graphics. Battery operated, mystery action, tin with plastic. 15.75 in (40 cm). 1976. Scarcity: 5. $75-$125.

Train Sets

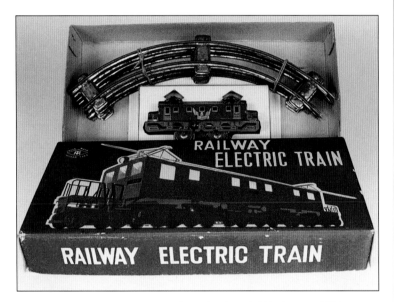

Railway Electric Train with circle of track. Windup, tin. 1952. Scarcity: 4. $100-$150.

#3453 **WR-2 Western Train Set**. Western style steam loco with tender and 2-rail sectional track. Windup, tin. 1966-1967. Scarcity: 4. $40-$80.

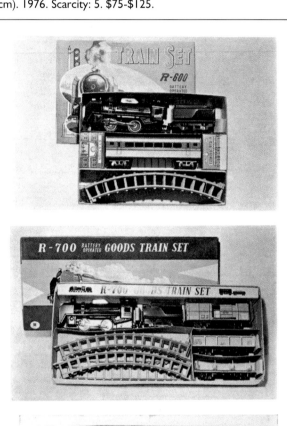

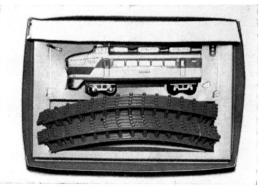

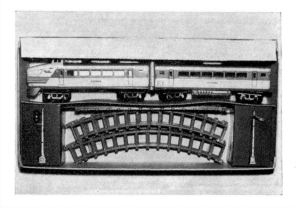

(top to bottom) #2000 **R-600 Train Set**. Steam engine and tender pulling passenger coach around 2-rail track. Battery operated, tin. 1960-1966. Scarcity: 4. $75-$125. #1998 **R-700 Goods Train Set**. Steam engine and tender pulling three freight cars around 2-rail track. Battery operated, tin. 1960-1966. Scarcity: 4. $100-$175. #3016 **Limited Express Kodama #1**. Kodama Express locomotive that circles 2-rail track. Battery operated, tin. 1961-1966. Scarcity: 3. $60-$100. #3017 **Limited Express Kodama #2**. Kodama Express locomotive and 'B' unit that circles 2-rail track. Battery operated, tin. 1961-1966. Scarcity: 3. $100-$150.

Train Sets 229

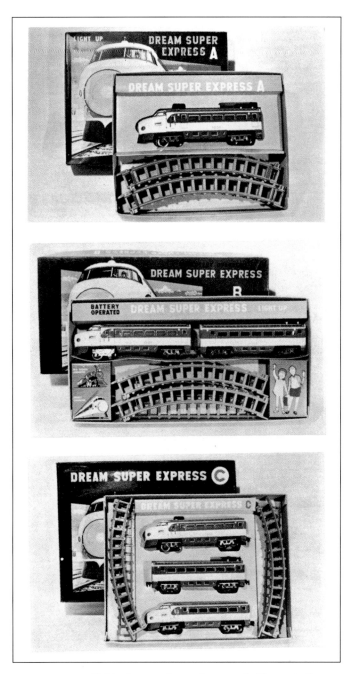

(top to bottom) #3186 **R-501 Smoky Train Set**. Steam engine and tender with smoke circle 2-rail sectional track. Battery operated, tin. 1962-1968. Scarcity: 2. $50-$100. #3228 **R-30 Central Train Set**. Steam loco and tender with passenger coach and 2-rail track. Windup, tin. 1962-1968. Scarcity: 4. $60-$100. #3274 **Train Set R-32**. Steam loco and tender with station platform, two passenger coaches, and 2-rail track. Windup, tin. 1963-1968. Scarcity: 5. $75-$125. #3322 **Smoking Train Set R-502**. Steam train with smoke. Set includes tender, coach, signals, and 2-rail sectional track. Battery operated, tin. 1964-1968. Scarcity: 4. $75-$125.

(top to bottom) #3276 **Dream Super Express A**. Electric bullet train with headlight and 2-rail sectional track. Battery operated, tin. 1963-1966. Scarcity: 4. $75-$125. #3277 **Dream Super Express B Light-Up**. Two car electric bullet train with headlight and 2-rail sectional track. Battery operated, tin. 1963-1966. Scarcity: 4. $100-$150. #3278 **Dream Super Express C**. Three car electric bullet train with headlight and 2-rail sectional track. Battery operated, tin. 1963-1966. Scarcity: 5. $125-$200.

#3456 **WR-5 Western Train Set**. Western style steam loco with tender, two coaches, tunnel, and 2-rail sectional track. Windup, tin. 1966-1968. Scarcity: 4. $60-$100.

230 Trains

#3484 **Casey Jones Train Set**. Western locomotive 1441 with lithographed passenger car that runs on 2-rail circle of track. Battery operated, tin. 1968-1971. Scarcity: 3. $50-$100.

#3630 **Wild West Train Set** with lithographed coach, whistle, sound, and sectional track. Battery operated, tin. 1968-1969. Scarcity: 3. $60-$100.

#3720 **R-200 Train Set** with smoking engine, tender, and sectional track. Battery operated, plastic and tin. 1969-1973. Scarcity: 3. $25-$40.

(top to bottom) #3515 **R-1500 Train Set** includes tender, coach, signals, and 2-rail sectional track. Battery operated, tin with plastic. 1967-1968. Scarcity: 4. $50-$100. #3516 **R-1600 Goods Train Set** includes tender, three freight cars, signals, and 2-rail sectional track. Battery operated, tin with plastic. 1967-1968. Scarcity: 4. $75-$125. #3517 **R-1700 Train Set** includes tender, two coaches, freight car, signals, and 2-rail sectional track. Battery operated, tin with plastic. 1967-1968. Scarcity: 5. $100-$150. #3541 **SE-1 Express Set** with bullet train 1001. Battery operated, tin with plastic. 1967-1969. Scarcity: 4. $50-$100.

Train Sets   231

#3721 **R-210 Train Set (Passenger)** with smoking engine, tender, passenger coach, two signals, and sectional track. Battery operated, plastic and tin. 1969-1973. Scarcity: 3. $40-$60.

#3722 **R-220 Train Set (Freight)** with smoking engine, tender, two freight cars, two signals, and sectional track. Battery operated, plastic and tin. 1969-1973. Scarcity: 3. $50-$75.

Train Sets – Not Pictured

#1736 **Train Set No. 1000** steam engine, tender and signals with oval 2-rail track. Battery operated, tin. 1957-1958. Scarcity: 4. $75-$150.

#1892 **R-11 Electric Rail Set** Express R-11 electric rail car with single pantograph on circular track platform. Windup, tin. 1958. Scarcity: 5. $150-$200.

#3018 **Limited Express Kodama #3**. Two Kodama Express locomotives and one 'B' unit that circles 2-rail track. Battery operated, tin. 1961-1966. Scarcity: 4. $125-$200.

#3065 **R610 Train Set**. Steam train with tender, freight car, and two passenger coaches. Battery operated, tin. 1961-1966. Scarcity: 5. $125-$200.

#3190 **Dream Super Express #2**. Bullet train, coach and track. Battery operated, tin. 1961-1964. Scarcity: 6. $125-$225.

#3259 **R-102 Train Set**. Bullet train, coach, and track. Windup, tin. 1963-1964. Scarcity: 6. $100-$150.

#3260 **R-103 Train Set**. Three car bullet train, signals and sectional track. Windup, tin. 1963-1964. Scarcity: 7. $150-$225.

#3268 **Dream Super Express #1**. Bullet train with sectional track. Battery operated, tin. 1963-1964. Scarcity: 6. $100-$175.

#3331 **R-801 Train Set**. Steam train with tender and 2-rail sectional track with ballast. Battery operated, tin. 1964-1966. Scarcity: 5. $60-$100.

#3332 **R-802 Train Set**. Steam train with tender, coach, signals, and 2-rail sectional track with ballast. Battery operated, tin. 1964-1966. Scarcity: 5. $75-$125.

#3333 **R-803 Train Set**. Steam train with tender, two freight cars, and 2-rail sectional track with ballast. Battery operated, tin. 1964-1966. Scarcity: 5. $100-$150.

#3369 **Dream Super Express with Bridge**. Two car electric bullet train with headlight, 2-rail sectional track, and bridge with elevated supports. Battery operated, tin. 1965-1967. Scarcity: 5. $125-$175.

#3454 **WR-3 Western Train Set**. Western style steam loco with tender, coach, and 2-rail sectional track. Windup, tin. 1966-1967. Scarcity: 4. $45-$85.

#3455 **WR-4 Western Train Set**. Western style steam loco with tender, two coaches, and 2-rail sectional track. Windup, tin. 1966-1967. Scarcity: 4. $50-$90.

#3542 **SE-2 Express Set** with two car bullet train 1001 and 1002. Battery operated, tin with plastic. 1967-1969. Scarcity: 4. $60-$125.

#3543 **SE-3 Express Set** with three car bullet train 1001, 1002, and 1003. Battery operated, tin with plastic. 1967-1969. Scarcity: 4. $75-$150.

#3550 **Smoking Train Set** with tender, signals, and plastic sectional track. Battery operated, tin with plastic. 1968. Scarcity: 6. $40-$60.

#3551 **Smoking Train Set** with tender, passenger coach, signals, and plastic sectional track. Battery operated, tin with plastic. 1968. Scarcity: 6. $50-$75.

#3731 **Smoking Train Set R-310** with smoking engine, tender, passenger coach, two signals, and plastic sectional track. Battery operated, plastic and tin. 1969-1970. Scarcity: 5. $40-$60.

#3772 **R-720 Smoky Train Set**. Smoking steam engine and tender circle sectional track. Battery operated, plastic. 1970-1971. Scarcity: 3. $30-$50.

#3804 **R-400 Play Train Set** with smoking engine, tender, passenger coach, and sectional track. Battery operated, plastic. 1970-1971. Scarcity: 3. $30-$50.

#3810 **Whistle Pioneer Express Set** with whistle, sound, and open freight wagon on sectional track. Battery operated, plastic with tin. 1970-1971. Scarcity: 4. $30-$50.

#3822 **Super Express Train Set**. Bullet train and coach with sectional plastic track. Battery operated, tin and plastic. 1971-1973. Scarcity: 5. $40-$75.

#3880 **R-800 Train Set** with smoking engine and tender on sectional plastic track. Battery operated, plastic. 1971-1973. Scarcity: 3. $25-$40.

#3881 **R-810 Train Set** with smoking engine, tender, and passenger car on sectional track. Battery operated, plastic and tin. 1971-1973. Scarcity: 3. $30-$50.

Trucks & Construction

Trucks

(top) **World's Express** with open hauler. Windup, tin. 8 in (20 cm). 1930s. Scarcity: 9. *Bill & Stevie Weart Collection.* $800-$1,200. (bottom) **Gasoline Truck** with articulated tanker. Windup, tin. 8 in (20 cm). 1930s. Scarcity: 9. *Bill & Stevie Weart Collection.* $800-$1,200.

#1729 **Cement Mixer** with rotating mixer. Battery operated, tin. 11.5 in (29 cm). 1957-1958. Scarcity: 6. $150-$250.

Mobilgas Oil Truck. Friction, tin. 8.5 in (22 cm). 1950s. Scarcity: 6. $125-$225.

Super Truck. Chevrolet stake body pickup. Friction, tin. 16 in (41 cm). 1950s. Scarcity: 8. $500-$800.

#3196 **Lucky Cement Mixer** with lithographed windows, rotating plastic mixing drum with balls. Battery operated, mystery action, tin with plastic. 12 in (30 cm). 1961-1964. Scarcity: 5. $125-$200.

#3252 **Dump Truck**. Chevrolet truck with dumping action and tin shovel. Friction, tin. 11.5 in (29 cm). 1962-1967. Scarcity: 7. $400-$700.

#3395 **Chevrolet Deluxe Truck – JNR**. Detailed Chevrolet truck with JNR cardboard container. Friction, tin. 11 in (28 cm). 1965-1968. Scarcity: 8. $600-$1,200.

#3284 **Chevrolet Deluxe Truck**. Detailed Chevrolet truck with fender mirrors. Friction, tin. 11 in (28 cm). 1964-1968. Scarcity: 8. $600-$1,200.

#3460 **Gasoline Trailer Truck**. Mobilgas tanker truck. Friction, tin. 17 in (43 cm). 1966-1967. Scarcity: 5. $150-$225.

#3380 **Overland Transport Express**. Green truck and trailer with blinking light and realistic horn sound. Battery operated, mystery action, tin. 17 in (43 cm). 1965-1967. Scarcity: 5. $150-$250.

Trucks – Not Pictured

#1604 Milk Trailer. M&K Sweet Milk open top hauler. Friction, tin. 12 in (30 cm). 1956-1960. Scarcity: 4. $100-$200.

#1625 Fruits Trailer. Fresh Oranges open top hauler. Friction, tin. 9.25 in (23 cm). 1956-1960. Scarcity: 3. $75-$150.

#1672 Kiddy Truck. Open type hauler with Kiddy Truck sign on sides. Friction, tin. 12.5 in (32 cm). 1956-1960. Scarcity: 3. $75-$125.

#1730 Dump Truck with Light with hydraulic dump action and light. Battery operated, tin. 11.5 in (29 cm). 1957-1958. Scarcity: 6. $125-$225.

#1785 Truck with low sided bed. Friction, tin. 9 in (23 cm). 1957-1958. Scarcity: 3. $40-$75.

#1787 Dump Car with raising dump bed. Friction, tin. 9.5 in (24 cm). 1957-1958. Scarcity: 3. $60-$100.

#1823 Dump Truck with crank raised dump bed. Friction, tin. 11 in (28 cm). 1958-1960. Scarcity: 3. $75-$125.

#1902 Super Mixer with plastic lite-up mixing drum. Battery operated, tin and plastic. 12 in (30 cm). 1958-1960. Scarcity: 5. $100-$175.

#1915 Communication Truck with dual Morse code units. Battery operated, tin. 12.5 in (32 cm). 1958. Scarcity: 6. $125-$200.

#3129 Lucky Mixer (animal truck) with animal graphics and rotating plastic mixing drum with balls. Battery operated, non-stop, tin. 12 in (30 cm). 1961-1962. Scarcity: 4. $100-$175.

#3282 Chevrolet Deluxe Truck with Cover. Detailed Chevrolet truck with vinyl top and Japanese writing. Friction, tin. 11 in (28 cm). 1964. Scarcity: 8. $600-$1,200.

#3356 Transport Express. Red trailer truck. Friction, tin. 17 in (43 cm). 1965-1967. Scarcity: 4. $125-$200.

Fire Trucks

#1935 Old Fashioned Fire Engine Carrying Free Flying Ball with driver and ball that floats above boiler stack. Battery operated, mystery action, tin. 10.5 in (27 cm). 1959. Scarcity: 5. $100-$200.

#3307 Old Fashioned Fire Engine with bell sound and flashing light. Battery operated, mystery action, tin. 12.5 in (32 cm). 1964-1967. Scarcity: 3. $100-$175.

Fire Car with four lithographed firemen, driver, bell, and extending ladder. Friction, tin. 5.5 in (14 cm). 1950s. Scarcity: 5. $75-$125.

Old Smoky Joe No.2 with fireman driver, sparking boiler, and siren sound. Friction, tin. 5.5 in (14 cm). 1950s. Scarcity: 4. $75-$125.

#3439 Hook-and-Ladder Truck with 4 position lever control for extending ladder and direction, flashing light, and sound. R/C battery operated, tin. 13.75 in (35 cm). 1966-1969. Scarcity: 5. $150-$225.

Fire Trucks 235

#3495 **B-Z Hook-and-Ladder Truck** with manual extending ladder, firemen lithographed on windows, flashing lights, and sound. Battery operated, mystery action, tin with plastic. 13.75 in (35 cm). 1967-1969. Scarcity: 6. $150-$250.

#3534 **Comical Old Fire Engine** with horn and engine sound, flashing lights, and shaking front end with comic vinyl driver. Battery operated, mystery action, tin with vinyl. 13 in (33 cm). 1967-1968. Scarcity: 7. $150-$250.

#3685 **Fire Engine No. 1** with driver that moves from side to side, bell, siren, and flashing light. Battery operated, mystery action, tin. 16 in (41 cm). 1969-1973. Scarcity: 4. $125-$175.

#3728 **Fire Engine w/ Siren** with large truck mounted crank siren and ladders. Friction, tin with plastic. 11 in (28 cm). 1969-1973. Scarcity: 4. $100-$150.

#3675 **Old Fashioned Fire Engine w/ Ring Smoke** with vinyl driver, bell sound, and smoking boiler. Battery operated, mystery action, tin with vinyl. 10.75 in (27 cm). 1968-1970. Scarcity: 5. $100-$150.

#3759 **Siren Fire Engine** with plastic fire driver and rear standing fireman who cranks siren. Battery operated, mystery action, tin. 10.75 in (27 cm). 1969-1973. Scarcity: 4. $125-$200.

236 Trucks & Construction

#3815 **Fire Engine** with push button extending ladder. Friction, tin with plastic. 10.5 in (27 cm). 1970-1973. Scarcity: 4. $100-$175.

#4070 **Snorkel Fire Engine** with bell, sound, and rising and turning snorkel and fire figure. Battery operated, stop and go, tin. 13.5 in (34 cm). 1972-1977. Scarcity: 4. $100-$175.

#3849 **B/Z Hook-and-Ladder Truck** with forward/reverse, sound, flashing light, and extending ladder by remote control. Battery operated, tin. 13.75 in (35 cm). 1970-1973. Scarcity: 5. $125-$225.

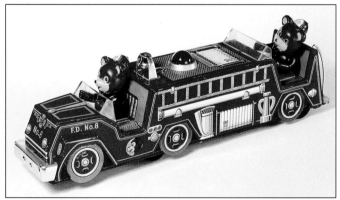

#4184 **Bear Squad Fire Engine** with flashing light, bell, front and rear driver. Battery operated, mystery action, tin with plastic. 16 in (41 cm). 1974-1975. Scarcity: 5. $75-$150.

Fire Trucks – Not Pictured

#1786A **Fire Engine Car** with extending ladder and four full-figured firemen. Friction, tin. 10 in (25 cm). 1957-1958. Scarcity: 6. $125-$200.

#1786B **Fire Engine Car** with extending ladder and two full-figured firemen. Friction, tin. 10 in (25 cm). 1960-1964. Scarcity: 5. $100-$175.

#3851 **R/C Hook-and-Ladder Truck** with siren sound, flashing light, and extending ladder with fireman. R/C battery operated, mystery action, tin. 13.75 in (35 cm). 1970-1973. Scarcity: 5. $100-$175.

Cranes   237

Bulldozers

#1610A **K-55 Electric Tractor** with rubber treads. Battery operated, tin. 7 in (18 cm). 1956. Scarcity: 4. $100-$175.

Bulldozers – Not Pictured

#1610B **K-55 Tractor with Trailer**. 7-inch tractor with rubber treads, driver, and trailer. Battery operated, tin. 12 in (30 cm). 1956-1960. Scarcity: 5. $125-$200.

#1674 **Tractor No.156**, "Excavating Construction" tractor with blade and tin driver. Friction, tin. 6.5 in (17 cm). 1956-1960. Scarcity: 4. $40-$80.

#1687 **Bulldozer No.25** with tin driver, blade, and rubber treads. Battery operated, tin. 8 in (20 cm). 1957-1960. Scarcity: 4. $75-$125.

#1845 **Bulldozer No.18 with Shovel** with tin driver, raising front loader, and rubber treads. R/C battery operated, tin. 10 in (25 cm). 1958-1960. Scarcity: 5. $125-$250.

High Power Crane with rotating tower and lifting bucket. Battery operated, tin. 1960s. Scarcity: 5. $200-$300.

Cranes

#3457 **Mighty King Mobile Crane** with rubber treads and 4 position lever control for bucket and direction. R/C battery operated, tin. 12 in (30 cm). 1966-1969. Scarcity: 5. $175-$275.

#4145 **R/C Crane** with adjustable direction, realistic treads, and lifting action. R/C battery operated, tin and plastic. 12 in (30 cm). 1973. Scarcity: 4. $75-$125.

Forklifts

#1685 **Fork Lift Truck** with tin driver. Battery operated, tin. 11.5 in (29 cm). 1957-1960. Scarcity: 5. $150-$225.

#1842 **Fork Lift Truck w/Piston Action**. S1002 with raising lift and visible moving pistons. R/C battery operated, tin. 11 in (28 cm). 1958-1960. Scarcity: 5. $150-$225.

#4078 **Forklift Truck**. HP-4078 with vinyl driver, forward/reverse, and up/down lever control. Battery operated, tin and plastic. 12 in (30 cm). 1972-1975. Scarcity: 4. $75-$125.

#1775 **B-Z Fork Lift Truck** with tin driver. Battery operated, mystery action, tin. 10 in (25 cm). 1957-1958. Scarcity: 6. $150-$350.

#1728 **B-Z Porter** with tin driver and three pieces of luggage. Battery operated, tin. 8 in (20 cm). 1957-1958. Scarcity: 5. $150-$250.

#4334 **R/C Forklift Truck**. MT-01 with vinyl driver and 4 function remote for forward/reverse and up down movement. R/C battery operated, plastic with tin. 11.5 in (29 cm). 1975-1980. Scarcity: 3. $50-$100.

#3294 **Road Roller** number 507 with canopy, tin driver, and lights. Battery operated, tin. 9.5 in (24 cm). 1964-1967. Scarcity: 4. $125-$250.

Road Rollers

#1890 **Bubble Road Roller**. Western with bubbles blowing from plastic stack. Battery operated, mystery action, tin with plastic. 9.5 in (24 cm). 1958-1960. Scarcity: 4. $100-$200.

#3348 **New Road Roller/Steam Roller** moves back and forth with blinking light and turning driver's head. Battery operated, tin with vinyl. 9.5 in (24 cm). 1964-1969. Scarcity: 4. $125-$200.

Other Collectible Toys

Some of the toys shown in this section did not fit easily into other sections because of the mixed subjects represented. There are also a variety of newer collectible toy categories represented. They include Comic Toys, Counter Toys (on top of the toy counters in bulk packaging), Gobblers, Pullback Toys, windup Walkers and Water Toys.

Miscellaneous

Mechanical Super Highway. Cars move on belt across dimensional panorama. Windup, tin. 6.5 in (17 cm). 1952. Scarcity: 7. *Courtesy of Barbara Moran.* $900-$1,500.

Shin-Manshu. Two cars featuring Chinese and Japanese occupants travel up and down inclined platform, depicting the New Manchuria under Japanese control. Windup, tin. 1930s. Scarcity: 10. *Courtesy of Ray Rohr.* $8,000-$10,000.

Aquaworld picture frame with moving fish in water filled tank. Battery operated, plastic. 9.5 in (24 cm). 1996. Scarcity: 1. $20-$35.

Mechanical Panorama Coast Line. Trains move on belt across dimensional panorama with moving boat. Windup, tin. 6.5 in (17 cm). 1950. Scarcity: 8. $1,000-$1,800.

Comic Toys 241

#1777 **Tower Bridge**. Tram travels back and forth on platform bridge with London towers. Battery operated, tin. 41.5 in (105 cm). 1957-1958. Scarcity: 8. $250-$450.

Miscellaneous – Not Pictured

Main Street with vehicles pulled around platform base. Diorama of street includes large traffic light, buildings, and tunnel. Windup, tin. 15 in (38 cm). 1930s. Scarcity: 9. $1,200-$2,000.

#1818 **Circling Monorail**. Circular platform with elevated monorail and flying plane traveling in a circle. Windup, tin. 5.5 in (14 cm). 1958-1961. Scarcity: 7. $125-$200.

#3067 **Lucky Express** train goes through tunnels and around platform base while helicopter circles overhead. Windup, tin. 6 in (15 cm). 1960-1962. Scarcity: 6. $75-$150.

#3068 **Snow Mobile**. Penguin snowmobile with antenna lever control. Battery operated, tin. 9 in (23 cm). 1960-1962. Scarcity: 6. $125-$225.

#3003 **Comic Car – Fire Engine** with face, bell, eye-dropper hose nozzle, and animal lithography. Friction, tin and plastic. 7 in (18 cm). 1960-1964. Scarcity: 3. $40-$60.

Comic Toys

#4052 **Comic Coffee Pot** with rattling lid, smoke, and moving eyes. Battery operated, mystery action, tin and plastic. 9 in (23 cm). 1972-1974. Scarcity: 4. $75-$150.

#3002 **Comic Car – Mixer**. Truck with face, clear mixer, and animal lithography. Friction, tin and plastic. 7 in (18 cm). 1960-1964. Scarcity: 3. $40-$60.

#4052s **Comic Coffee Pot** mock-up sample. Battery operated, mystery action, tin and plastic. 9 in (23 cm). 1972-1974. Scarcity: 10.

Other Collectible Toys

#4053 **Comic Cooking Pot** with rattling lid, smoke, and moving eyes. Battery operated, mystery action, tin and plastic. 9 in (23 cm). 1972-1974. Scarcity: 4. $75-$150.

#4054s **Comic Toaster** mock-up sample. Battery operated, mystery action, tin and plastic. 7 in (18 cm). 1972. Scarcity: 10.

Comic Toys – Not Pictured

#3004 **Comic Car – Circus**. Truck with face, revolving spoke wheel, and animal lithography. Friction, tin and plastic. 7 in (18 cm). 1960-1964. Scarcity: 3. $40-$60.

#3856 **Comical Clock** with shaking comical face, feet and hands. Windup, tin with plastic. 4 in (10 cm). 1970-1972. Scarcity: 3. $20-$40.

Counter Toys

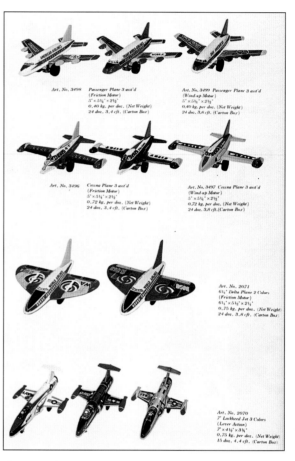

Counter Toys - Aircraft. Windup and friction, tin and plastic. 1974-1980s. Scarcity: 2. $20-$35.

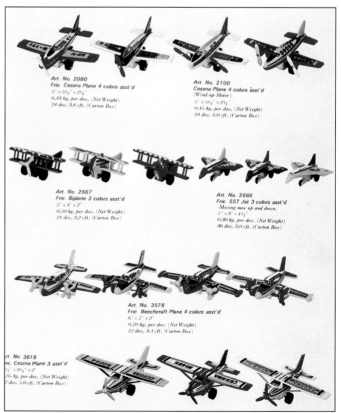

Counter Toys 243

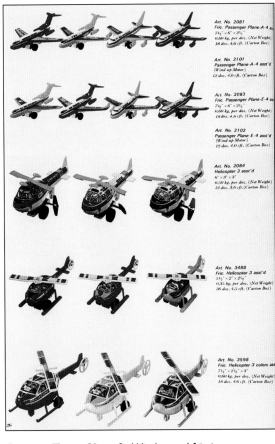

Counter Toys - Aircraft. Windup and friction, tin and plastic. 1974-1980s. Scarcity: 2. $20-$35.

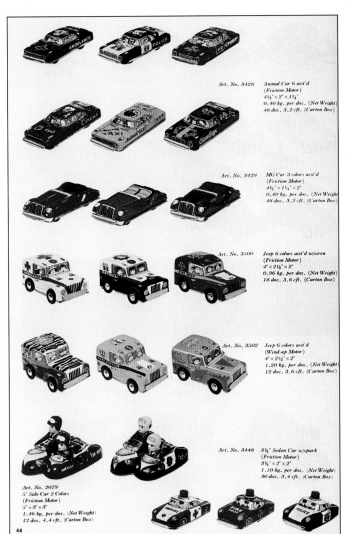

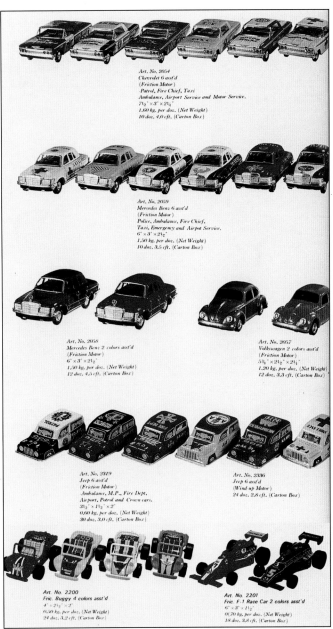

Counter Toys - Vehicles. Windup and friction, tin and plastic. 1974-1980s. Scarcity: 3. $15-$50.

Counter Toys - Vehicles. Windup and friction, tin and plastic. 1974-1980s. Scarcity: 2. $15-$25.

244   Other Collectible Toys

Counter Toys - Vehicles. Windup and friction, tin and plastic. 1974-1980s. Scarcity: 2. $15-$25.

Gobblers

#4481 **Swimming Hippo, Fish & Frog** swim and swallow little fish on string. Windup, plastic. 5.5 in (14 cm). 1976-1980s. Scarcity: 2. $10-$20.

#4541 **Gulping Shark** swallows little fish on string. Windup, plastic. 6.5 in (17 cm). 1977-1980s. Scarcity: 3. $15-$25.

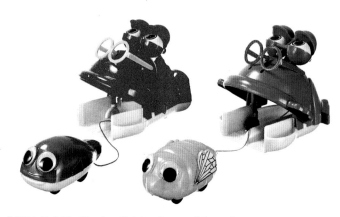

#4561 **Gobbly Gooks.** Gulping frog or fish swallows little fish on string. Windup, plastic. 5 in (13 cm). 1977-1980s. Scarcity: 3. $15-$25.

#4586 **Gulping Flying Fish, Whale & Dog** swim and swallow little fish or bone on string. Windup, plastic. 6.5 in (17 cm). 1977-1980. Scarcity: 3. $15-$25.

#4590 **Gulping Alligator, Frog & Dolphin** tread water and swallow little fish or toys on string. Windup, plastic. 5 in (13 cm). 1977-1980. Scarcity: 3. $10-$20.

Gobblers   245

#4618 **Safari Jeep** chasing and capturing rhinoceros. Windup, plastic. 9 in (23 cm). 1978-1980. Scarcity: 4. $25-$40.

#4675 **Chasing Police Car** chasing and capturing a robber. Windup, plastic with tin. 6 in (15 cm). 1978-1980. Scarcity: 4. $30-$50.

#4629 **Hungry Cat & Dog** chases mouse or sausage on string and swallows them. Windup, plastic. 5 in (13 cm). 1978-1980. Scarcity: 4. $15-$25.

(left to right) #4694 **Swimming Big Mouth Fish**. Gulping fish swallows little fish on string. Windup, plastic. 4 in (10 cm). 1979-1980s. Scarcity: 2. $10-$15. #4695 **Big Mouth Elephant**. Gulping elephant swallows peanut on string. Windup, plastic. 4 in (10 cm). 1979-1980s. Scarcity: 2. $10-$15. #4696 **Big Mouth Hippo** Gulping hippo swallows little fish on string. Windup, plastic. 4 in (10 cm). 1979-1980s. Scarcity: 2. $10-$15.

#4645 **Money Gobblers** swallow play coin on string while walking. Windup, plastic. 6 in (15 cm). 1978-1980s. Scarcity: 2. $10-$20.

Other Collectible Toys

#4809 **Disney Swimming Whale**. Mickey Mouse sits on top of gulping whale. Windup, plastic. 1981-1980s. Scarcity: 4. $20-$30.

#4567 **Pull Back Palm Pet Trains,** assorted trains – steam, electric, and express. Windup, pull back, plastic. 4 in (10 cm). 1978. Scarcity: 5. $15-$30.

#4642 **Pull Back Palm Pet Planes,** assorted planes 747, L1011 or DC10. Windup, pull back, plastic. 4 in (10 cm). 1978-1980s. Scarcity: 3. $10-$15.

Pullback Toys

#4345 **Pull Back Palm Pet Series,** assorted vehicles – fire, police, and rally car. Friction, pull back, plastic. 2.5 in (6 cm). 1975-1978. Scarcity: 2. $10-$15.

#4509 **Pull Back Palm Pet Trucks**. Container truck, gasoline tanker, and garbage truck. Windup, pull back, plastic. 4 in (10 cm). 1977-1978. Scarcity: 4. $10-$20.

#4453 **Pull Back Palm Pet Series,** assorted vehicles – cycle, helicopter, and police cycle. Windup, pull back, plastic. 4 in (10 cm). 1976-1978. Scarcity: 3. $10-$15.

Walkers

#2674 **Robot Walkers** produced in red, silver, or gold. Windup, plastic. 5 in (13 cm). 1979-1980s. Scarcity: 4. $25-$50.

Art. No. 4393 Art. No. 4394 Art. No. 4395

(left to right) #4393 **Walking Droopy** with walking action. Windup, plastic with tin base. 4 in (10 cm). 1976-1980s. Scarcity: 4. $30-$40. #4394 **Walking Tom** with walking action. Windup, plastic with tin base. 4 in (10 cm). 1976-1980s. Scarcity: 4. $30-$40. #4395 **Walking Jerry** with walking action. Windup, plastic with tin base. 4 in (10 cm). 1976-1980s. Scarcity: 4. $30-$40.

Walkers 247

(left to right) #4420 **Walking Lippy The Lion** with walking action. Windup, plastic with tin base. 4 in (10 cm). 1976-1977. Scarcity: 7. $30-$45. #4419 **Walking Wally Gator** with walking action. Windup, plastic with tin base. 4 in (10 cm). 1976-1977. Scarcity: 7. $30-$45. #4418 **Walking Yogi Bear** with walking action. Windup, plastic with tin base. 4 in (10 cm). 1976-1977. Scarcity: 7. $30-$45.

#4516 **Walking Vegetables Assortment (6)**. Tomato, pepper, onion, corn, cucumber, or mushroom with walking action. Windup, plastic with tin base. 4 in (10 cm). 1977-1980s. Scarcity: 3. $15-$25.

Art. No. 4437   Art. No. 4438   Art. No. 4557

(left to right) #4437 **Walking Donald Duck** with walking action. Windup, plastic with tin base. 4 in (10 cm). 1976-1980s. Scarcity: 4. $30-$40. #4438 **Walking Mickey Mouse** with walking action. Windup, plastic with tin base. 4 in (10 cm). 1976-1980s. Scarcity: 4. $30-$40. #4557 **Walking Woody Woodpecker** with comical walking action. Windup, plastic with tin base. 4.25 in (11 cm). 1977-1979. Scarcity: 5. $30-$45.

#4558 **Walking Topo Gigio** with comical walking action. Windup, plastic with tin base. 4 in (10 cm). 1977. Scarcity: 8. $50-$75.

#4439 **Walking Robin Hood** with walking action. Windup, plastic with tin base. 4 in (10 cm). 1976-1977. Scarcity: 7. $40-$60.

#4573 **Walking Big Droopy** with walking action. Windup, plastic with tin base. 5 in (13 cm). 1977-1978. Scarcity: 6. $30-$50.

248   *Other Collectible Toys*

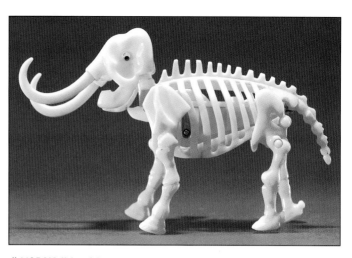

#4611 **Walking Penguin** with walking action. Windup, plastic with tin base. 4.25 in (11 cm). 1977. Scarcity: 5. $20-$40.

#4625 **Walking Mammoth** walking skeletal system. Windup, plastic. 10 in (25 cm). 1978-1980s. Scarcity: 4. $30-$50.

(left) #4619 **Pose Monkey**. Posing walking monkey with long arms. Windup, plastic with plush and tin. 4.25 in (11 cm). 1978-1979. Scarcity: 4. $10-$25. (right) #4612 **Walking Cook** with walking action. Windup, plastic with tin base. 4 in (10 cm). 1977-1980s. Scarcity: 3. $15-$30.

#4626 **Walking Comical Fruits**. Walking strawberry, pineapple and melon. Windup, plastic with tin. 4 in (10 cm). 1978-1980. Scarcity: 3. $20-$30.

#4624 **Walking Dinosaur** walking skeletal system. Windup, plastic. 10 in (25 cm). 1978. Scarcity: 7. $50-$75.

Walkers 249

#4649 **Walking Champion**. Kendo, baseball, and football walkers. Windup, plastic with tin. 4 in (10 cm). 1978. Scarcity: 4. $25-$35.

#4691/3 **Lovely Kisses Family**. Three walking character figures. Windup, plastic. 4 in (10 cm). 1979-1980s. Scarcity: 4. $15-$25.

#4701 **Walking Games**. Lucky Point, Gunfight, and Blackjack pinball game walkers. Windup, plastic. 6.75 in (17 cm). 1979. Scarcity: 6. $20-$40.

#4870 **Walking Duck** shakes head and quacks while walking. Windup, plastic. 5.75 in (15 cm). 1981-1980s. Scarcity: 3. $10-$20.

#4727 **Walking Family Panda Bears**. Three assorted walking panda families. Windup, plastic. 4 in (10 cm). 1979-1980s. Scarcity: 3. $15-$25.

Walkers – Not Pictured

#4863 **Mr. Slugger** walking baseball player with bat. Windup, plastic. 3.5 in (9 cm). 1981-1980s. Scarcity: 3. $10-$20.

Water Toys

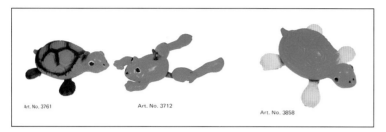

(left to right) **#3761 Tortoise The Swimmer** with water swimming action. Windup, plastic. 6.5 in (17 cm). 1969-1974. Scarcity: 1. $10-$20. **#3712 Frog The Swimmer** with water swimming action. Windup, plastic. 7.5 in (19 cm). 1969-1973. Scarcity: 2. $10-$20. **#3858 Swimming The Tortoise** with water swimming action. Windup, plastic. 9 in (23 cm). 1970-1973. Scarcity: 2. $10-$20.

#3967 Swimming The Duckling with water swimming action. Windup, plastic. 4.5 in (11 cm). 1971-1975. Scarcity: 2. $15-$25.

#3792 Swimming The Alligator with water swimming action. Windup, plastic. 9 in (23 cm). 1970-1973. Scarcity: 2. $10-$15.

#7252 Swimming the Lobster with water swimming action. Windup and friction, plastic. 9.5 in (24 cm). 1973-1974. Scarcity: 3. $20-$30.

#3990 Rowing Boat water toy with boy rowing oars. Windup, plastic with tin. 7 in (18 cm). 1972-1975. Scarcity: 3. $25-$50.

Water Toys – Not Pictured

#4099 Swimming the Turtle with water swimming action. Windup, plastic. 7.5 in (19 cm). 1973-1974. Scarcity: 3. $15-$25.

#7198 Swimming the Dragon with water swimming action. Windup, plastic. 9 in (23 cm). 1973. Scarcity: 4. $25-$35.

Ads and Catalogs

The following pages contain miscellaneous Masudaya trade journal ads and catalog pages.

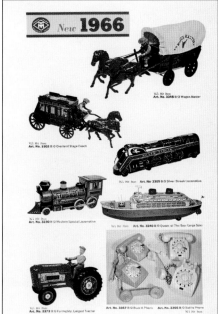

New items, 1966.

252 Ads and Catalogs

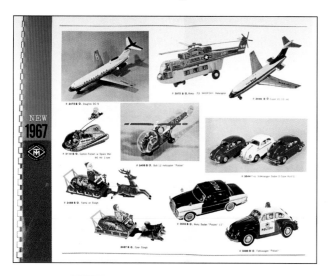

1969 catalog cover.

New items, 1967.

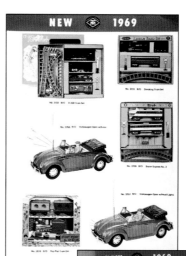

New items, 1969.

New items, 1968.

Ads and Catalogs 253

New items, 1970.

1973 catalog page.

1972 catalog page.

1973 catalog page.

1980s birds in cages.

Ads and Catalogs

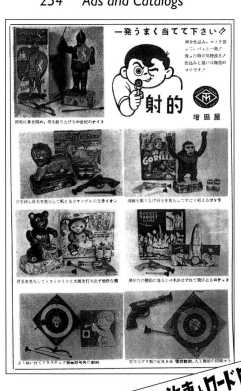

August 1958.

October 1962.

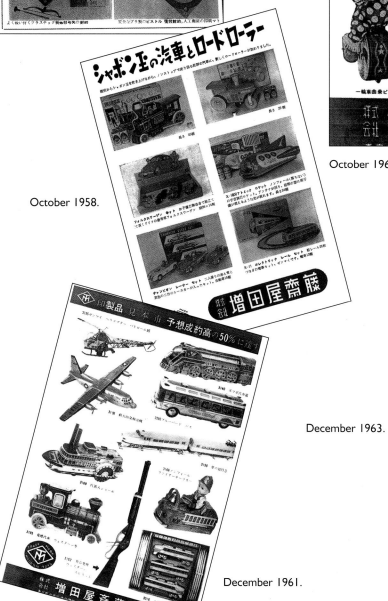

October 1958.

December 1961.

December 1963.

Ads and Catalogs   255

February 1964.

November 1964.

September 1965.

September 1965.

Bibliography and Resources

Gallagher, William C. *Japanese Toys - Amusing Playthings From The Past*. Atglen, Pennsylvania: Schiffer Publishing Ltd., 2000.

Gould, David C., and Donna Crevar-Donaldson. *Occupied Japan Toys*. Gas City, Indiana: L-W Book Sales, 1993

Japan Toy International Trade Fair Association. *1st Japan International Toy Fair Catalog*. 1962.

Japan Toys Museum Foundation, Shigeru Mozuka, Manager; Mitsuo Tsukuda, Owner and Founder. Tokyo, Japan.

Kaminaga, Eiji. CEO, Marusan Toys Inc. Conversations with the author. 2000-2004.

Kelley, Dale. *Antique Toy World* magazine. 2000-2004.

Kitahara, Teruhisa. *Cars Tin Toy Dreams*. San Francisco: Chronicle Books, 1985.

Kitahara, Teruhisa. *Robots Tin Toy Dreams*. San Francisco: Chronicle Books, 1985.

Kitahara, Teruhisa. *Wind-Ups Tin Toy Dreams*. San Francisco: Chronicle Books, 1985.

Kitahara, Teruhisa. *Teruhisa Kitahara Collection Tin Toy Box Graphic Arts* (in Japanese). Jitsugyo No Nihon Sya, Ltd., 1999.

Machii, Mikio. Director, Iwaya Corporation. Conversations with the author. Tokyo, Japan 2000-2004.

Marsella, Anthony R. *Toys From Occupied Japan*. Atglen, Pennsylvania: Schiffer Publishing Ltd., 1995.

Masudaya Toy Co. Ltd. *Catalogs 1958-1980*. Tokyo, Japan.

Morita, Kinya. *Robot & Space Toy's Collection*. Tokyo, Japan: World Mook 242, 2000.

Saito, Hank. Director, Masudaya Corporation. Conversations with the author. 2000-2004.

Smith, Herb. *Smith House Toys 1998 Price Guide*. Herb Smith, 1998.

Smith, Ron, and William C. Gallagher. *The Big Book of Tin Toy Cars - Passenger, Sports and Concept Vehicles*. Atglen, Pennsylvania: Schiffer Publishing Ltd., 2004.

Smith, Ron, and William C. Gallagher. *The Big Book of Tin Toy Cars - Commercial and Racing Vehicles*. Atglen, Pennsylvania: Schiffer Publishing Ltd., 2004.

Super 7 Media, Inc. *Super 7 Magazine*, Volume 1 Issues 1-4, 2003 and Volume 2 Issues 1-3, 2004.

Takayama, Toyoji. *Nostalgic Tin Toys vol. 21, Character Toys*. Kyoto: Kyoto Shoin, 1989.

The Tin Toy Museum and collection of Toyoji Takayama, Kyoto, Japan.

Tin Toy Museum, Teruhisa Kitahara. Yokohama, Japan.

Udagawa, Yoshio. Chairman of the Board, Toplay (T.P.S.) Ltd. Conversations with the author. Tokyo, Japan. 1998-1999.